Praise for *Pets on the Couch*

"Dr. Dodman presents a captivating look at the world of animals, who are more like ourselves than we know. Nick tells priceless stories of pets and the psychological and biological causes of their problems, which are all too human. Fun, informative, and surprising."

—JOHN J. RATEY, MD, author of *Spark* and coauthor of *Driven to Distraction*

"How many of us have wished that our veterinarian was our doctor? To know Dr. Nicholas Dodman is to wish he was your therapist. Dr. Dodman teaches us that when we underestimate animals, we underestimate all that can be done for them—and we most definitely underestimate ourselves. Dr. Dodman has never accepted the limits of what is 'known' and has again and again discovered what is possible."

—TRACEY STEWART, author of *Do Unto Animals* and editor-in-chief of the website Moomah

"Haven't all animal lovers wished at least once to be given Dr. Dolittle's power to converse with our four-footed friends? Dr. Dodman's delightful book, *Pets on the Couch*, might be the next best thing. His engaging and thoughtful observations, based both on neuroscience and a lifetime of experience, captivate the reader and open up a new window of understanding into our beloved equine and other animal companions."

—ELIZABETH LETTS, #1 New York Times bestselling author of *The Eighty-Dollar Champion* and *The Perfect Horse*

"Dr. Dodman cleverly utilizes his vast clinical experience to deliver a fascinating collection of insightful and essential behavioral tales. The delineation between human emotions and those of the animals in our lives has never been finer."

—**NICHOLAS TROUT, DVM, author of *Tell Me Where It Hurts***

"Nick Dodman has the best of both worlds—a brilliant mind and a kind heart. *Pets on the Couch* is a wonderful, heartbreaking, and inspiring journey through the animal mind. Follow Dodman as he tracks down the source of seemingly quirky behavior, the root of suffering, and the sometimes convoluted path to recovery. A must read for anyone who has a pet, who loves animals, or is interested in the mysterious workings of the brain. It's no wonder that Dodman is America's most beloved veterinarian."

—**BRIAN HARE, professor of cognitive neuroscience at Duke University and author of the *New York Times* bestseller *The Genius of Dogs***

By Nicholas H. Dodman

The Dog Who Loved Too Much

Dogs Behaving Badly

The Cat Who Cried for Help

If Only They Could Speak

The Well-Adjusted Dog

PETS ON THE
COUCH

Neurotic Dogs, Compulsive Cats, Anxious Birds,
and the New Science of Animal Psychiatry

NICHOLAS H. DODMAN, BVMS, DACVB

ATRIA PAPERBACK

New York   London   Toronto   Sydney   New Delhi

ATRIA
PAPERBACK

An Imprint of Simon & Schuster, Inc.
1230 Avenue of the Americas
New York, NY 10020

First Atria Paperback edition July 2017

ATRIA PAPERBACK and colophon are trademarks of Simon & Schuster, Inc.

For information about special discounts for bulk purchases, please contact Simon & Schuster Special Sales at 1-866-506-1949 or business@simonandschuster.com.

The Simon & Schuster Speakers Bureau can bring authors to your live event. For more information, or to book an event, contact the Simon & Schuster Speakers Bureau at 1-866-248-3049, or visit our website at www.simonspeakers.com.

Interior design by Amy Trombat

Manufactured in the United States of America

10 9 8 7 6 5 4 3 2 1

The Library of Congress has cataloged the hardcover edition as follows:
Names: Dodman, Nicholas H., author.
Title: Pets on the couch : neurotic dogs, compulsive cats, anxious birds, and the new science of animal psychiatry / by: Nicholas H. Dodman, BVMS, DACVB.
Description: First Atria Books hardcover edition. | New York : Atria Books, 2016. | Includes bibliographical references and index.
Identifiers: LCCN 2016019145 (print) | LCCN 2016025496 (ebook) | ISBN 9781476749020 (hardcover : alk. paper) | ISBN 9781476749037 (pbk. : alk. paper) | ISBN 9781476749044 (eBook)
Subjects: LCSH: Pets--Psychology. | Pets--Behavior. | Animal psychopathology.
Classification: LCC SF412.5 .D63 2016 (print) | LCC SF412.5 (ebook) | DDC 636.088/7--dc23
LC record available at https://lccn.loc.gov/2016019145

ISBN 978-1-4767-4902-0
ISBN 978-1-4767-4903-7 (pbk)
ISBN 978-1-4767-4904-4 (eBook)

For my mother, Gwen Dodman, who showed me
how to love and care for all animals

Empiricism takes no account of the soul.

—OLIVER SACKS

Behavior: that's what it's all about, really.

—MARGARET THATCHER, in personal conversation

Contents

1 **ONE**   The Dog Who Ate Wineglasses: When the Brain Short-Circuits

15 **TWO**   The Horse Who Chewed on Fences: Obsessive-Compulsive Disorder

27 **THREE**   The Dog Who Came Back from the War: Post-Traumatic Stress Disorder

41 **FOUR**   The Emotional Wounds of Rescued Animals

57 **FIVE**   The Dog Who Couldn't Stop Licking: Compulsive Disorders

81 **SIX**   The Dog Who Was Afraid of Puddles: Autism, Epilepsy, and Rage

99 **SEVEN**   The Horse Who Went "Harumph": Equine Tourette's Syndrome

115 **EIGHT**   The Dog Who Hated Surprises: The Many Faces of Aggressive Behavior

137 **NINE**   Two Spaniels and a Baby: Predation and Pharmacological Fixes

153 **TEN**   Animals Who Fear Too Much: Anxiety and Panic Disorders

173 **ELEVEN**   Dogs Who Hate Bugs and Storms: The Trouble with Phobias

187 **TWELVE**   Senile Dogs, Cats…and Cheetahs: Dementia and Alzheimer's Disease

199 **THIRTEEN**   The Beagle with ADHD: Behavioral Problems with Medical Roots

213 **FOURTEEN**   The Narcoleptic Horse: Night Terrors and Other Sleep Problems

223 **FIFTEEN**   The Listless Pet: The Thyroid and Anxiety, Aggression, and Mood

233 **SIXTEEN**   Blue Dogs and Cats: Depression and Mood Medicines

249 **EPILOGUE**   Hope for Us All

263 **Acknowledgments**

265 **Bibliography**

271 **Index**

PETS ON THE
COUCH

CHAPTER ONE

The Dog Who Ate Wineglasses

When the Brain Short-Circuits

> Some people talk to animals. Not many listen though.
> That's the problem.
>
> —A. A. MILNE

I sat in my office and watched a video of a male Golden retriever going insane.

The footage was murky because it was shot in a darkened house at night. The animal was sleeping peacefully on his blanket, the picture of domestic peace, until his muzzle began to twitch. In a gradual transformation reminiscent of Dr. Jekyll and Mr. Hyde, the twitching became more rapid, and suddenly, the beast jumped up and attacked his own bed.

The Golden's name was Comet, and the speed of his abrupt assault was worthy of his name. His targets varied, according to his owner. On the video, Comet attacked a blanket, snarling, snapping, and ripping the thing to shreds, but his owner told me that he also occasionally tore into his canine housemate, an English setter, sending the poor thing yowling in retreat. Somehow the violence appeared just as vicious when Comet's prey was an innocent patch of fabric.

I looked from the monitor to the dog who lay placidly at my feet. This was the monster I had just witnessed going berserk? During the day Comet was a gentle dog who would happily play-wrestle with the English setter and always lose, content to end up on his back on the ground with the setter lording it over him. At night, though, he became aggressive. Even though the attacks were over in a few seconds, they were scary to watch.

In my practice as a professor of veterinary science specializing in animal behavior, I had encountered such episodes before. A bull terrier would wake abruptly in a rage and attack a door. Another bull terrier would try to attack his master, though I had advised her to attach him to a metal leash that was long enough for him to move around but not so long that he could reach her while she was sleeping. More than once she woke up with the dog snarling and snapping inches from her face. She was extremely grateful for that metal leash!

The cause behind each unusual behavior in pets is a puzzle to be solved. Sometimes they're multiple. Sometimes they're biological. But sometimes biological causes create psychological problems. Comet the loving, playful sweetie pie became Comet the snarling nighttime werewolf. Bully the loving pet and Bully the berserker. Every owner of every dog has a secret knowledge, at times relished, at times feared, that there is an animal in the house who back in the mists of time was once a wild beast. But when a dog inexplicably reverts to its wolflike nature, can anything be done?

I diagnosed a neurological problem in Comet, a possible seizure disorder, so I prescribed meds that had anticonvulsant properties. With a combination of clonazepam, a Valium-type anticonvulsant, and phenobarbital, Comet's nighttime attacks reduced considerably.

For other dogs suffering similar issues, I used similar approaches. The metal-leashed bull terrier responded well to Prozac, which has antiepileptic as well as mood-stabilizing properties, and he stopped his nightly attack. Later I was able to keep him calm by switching him from Prozac to nighttime melatonin, which is also considered something of an anticonvulsant.

Let's stop a moment and take a deep breath. The preceding paragraph might strike you as worrisome. "Dr. Dodman, you propose to give my pet *what?*" Many pet owners prefer to hear about herbal remedies and nonpharmacological measures. And of course we often employ them, too. But I'm going to suggest that treating animals with human medications is not an instance of Big Pharma run amok. In the course of this book, I hope you will come to see the simple, practical truth. Modern medicines work. They alleviate suffering and save lives.

Such treatment approaches weren't arrived at by luck, and weren't random stabs at a cure, either. I knew how to address Comet's problem because, long before he came into my examining room, I had embarked upon a quest to understand such bizarre aggressive episodes. I delved into the available research on animals and had also reached across the species barrier to investigate the basis for sudden unexplained aggression in humans. When I put it all together, it seemed clear that there was a neurological trigger for what was happening.

The dogs weren't *choosing* to go berserk. Their attacks weren't the result of bad dreams. The afflicted animals were experiencing seizures, the result of misfires deep within the brain.

In my search of the available case histories I encountered many accounts of nocturnal seizures, and most were expressed as aggression. Why did such episodes occur only at night? Once again, the available case histories have long shown that sleep can promote seizures in people. It turns out that some natural mechanisms that inhibit behavior are switched off in deep sleep.

Back then, at the dawn of my career as an animal behaviorist, we didn't know much about the neural processes involved in this often upsetting and sometimes dangerous behavior. So I recruited fellow researchers to develop a way of addressing it. Our approach was remarkable in one respect: we proposed to treat animals with many of the same treatments that are successful in healing people.

On the one hand, our method might seem obvious. Human beings and canines are both vertebrates, both mammals. We are evolutionary relatives—distant relatives, but relatives all the same. Complex partial seizures are more common than grand mal seizures in people, so it made sense to assume that dogs had a high incidence of complex partial seizures, too. Why shouldn't we find common treatments of our shared pathologies?

But science doesn't assume. Science proves. Obvious as our hypothesis might be, there was a problem verifying it: How could we really tell when partial seizures were happening in dogs? People can describe experiences to their doctors, but animals obviously cannot. This probably explains why complex partial seizures remain undiagnosed by many veterinarians.

Despite the obvious success of the treatment regimens I was developing, many veterinary behaviorists were skeptical that seizure-based aggression existed. An eminent animal behaviorist, Dr. Ilana Reisner, who did her research for her doctoral thesis on springer spaniel aggression, stopped short of saying definitively the condition was seizure-based. One widely accepted textbook labeled sudden violent aggression in dogs as "idiopathic," a wonderful five-dollar word that simply means the cause is undetermined. It's what vets and doctors alike say when they are loathe to pronounce their least favorite three-word phrase: "I don't know."

For all this academic back-and-forth, animals were still coming to vets in distress. Successfully treating patients was one thing. Verifying the cause of their behavior was quite another.

A vital step along the way came when I saw Brock, a really big, 150-pound Chesapeake Bay retriever, whose aggression was completely out of control. The dog had put his owner in the hospital on more than one occasion. He was such a beast that he would even steal wineglasses and crush them in his jaws in front of his terrified owner. If the owner attempted to intervene, the dog would attack her viciously.

It was hard to understand how this much-bitten owner remained loyal to her dog, but love often wins out with pets. Fortunately, she brought Brock in to see me, and I in turn discussed the case with Dr. Klaus Miczek, a fellow Tufts professor, who agreed that partial seizures, or temporal lobe epilepsy, might be behind the savage attacks.

Dr. Miczek told me of a human equivalent called "episodic dyscontrol," a seizure-based condition that makes people respond violently and disproportionately to trivial triggers. Since this kind of condition existed in humans, it was conceivable that it could appear in dogs. It also gave us an encouraging sign for a possible treatment.

Klaus and I, together with veterinary neurologist, Dr. Johann Thalhammer, gave the dog an EEG, an electroencephalogram, which is used to detect abnormalities in electrical impulses in the brain. Sure enough, the results of Brock's EEG indicated a large complex spiking of electric impulses in the temporal lobe region, which in both dogs and people is considered "command central" of emotional control. The electrical activity detected in Brock's case was a sure sign of episodic dyscontrol or partial seizures.

Since a single case isn't enough to sustain a hypothesis, we examined two other dogs who had violent uncontrolled aggression and found that they also displayed similar abnormal electrical activity. All three dogs suffered bizarre mood changes right before their attacks began, then explosive aggression directed at people or objects.

The poor dogs! They weren't choosing to be aggressive. Their behavior wasn't purposeful. All three dogs responded well to a treatment of phenobarbital, a medicine also widely prescribed to humans in cases of epilepsy. This was not coincidence. What works in people turns out to work well in canines. All the owners of the treated dogs considered the improvement unqualified successes. They had saved their dogs.

So, happy endings all around. It's a treatment approach that I've used over and over throughout the years since, from Brock all the way forward to Comet. The pattern of positive outcomes points to a fundamental truth that I have spent my whole career attempting to fully understand. If I find behavior in an animal that I don't understand, I often look at similar behaviors in our own species to discover an effective method of treatment. Seizures in dogs? Why not look into the history of seizures in humans?

Despite publication of our evidence for the existence of episodic dyscontrol in dogs, however, some of my colleagues still refuse to accept it as possible. It is often hard to convince people that what they have believed for years is not true. I usually calculate that it takes about twenty years after publication of a new concept for it to enter the mainstream and become accepted. If that is the case, then the present time is just about right for that to happen.

In these specific instances—in the cases of the canine wineglass crusher, the metal-leash strainer, the fearsome blanket ripper—we just may win the battle to save animals' lives. But the wider war is ongoing. The real fight to save animals from destructive behavior—and from being euthanized or destroyed—is just beginning.

———

Are animals like us? Are we like them? Do they have intelligence? Do they have emotions? Can they suffer as we do? Most people who live with pets answer these questions with a rousing affirmative, as do veterinarians like me. Deep in their hearts, people who routinely work with livestock and farm animals would also agree—even though they might be loath to admit it, since their jobs may depend on ignoring such hard truths.

But there is one group that stubbornly resists acknowledging the validity of such propositions. Powerful sectors of the scientific community deny suppositions that seem perfectly obvious to lay people. The "real" scientists inform us that we must interpret animal behavior in mechanistic terms, rather than attributing what we see to higher brain functions.

Is your cat jealous? Does your dog exhibit empathy? Is a horse who continually paces her stall experiencing anxiety? No, no, and no, answer these stern arbiters of scientific purity. It's all reflex, not cognition, they say. Our pets are different from us, they say, and cannot share our thought processes.

Wary of appearing less rigorous than befits a scientist, the naysayer police protect the borders of the realm of *Homo sapiens*, turning away incursions from the wider animal kingdom. They are the champions of human exceptionalism.

So, okay. Maybe we just have a difference of opinion. Perhaps we *are* being soft-headed when we think Fluffy looks happy today when she looked sad yesterday. The average pet owner believes one way, the research scientist another. What does it really matter?

It might make no difference at all, except for one thing: animals are dying by the thousands because of obstinate, outmoded notions about their emotions and behavior. As a society, we have refused to make strides in understanding our pets because we are locked into old biases that limit vital avenues of research. As surely as the carriage horses who clop along the lanes of New York's Central Park, the scientific community wears blinders in its approach to animal behavior.

Meanwhile, "bad behavior" is the number one killer of pets in the United States.

What? How can that be? The leading killer of dogs and cats surely must be something like cancer, heart ailments, or simple old age. The toll of these diseases is, of course, high, but none can match the slaughter done in the name of bad behavior. It happens too often even to be remarkable. What is the answer when a dog turns aggressive and starts attacking people? When a cat becomes incontinent and fouls a new carpet? Off to the pound with it! Euthanasia is such a bland name for what actually occurs. A pity, but it has to be done.

But does it have to be done?

I've devoted my professional life to finding answers to why an animal's behavior goes awry. My method is simple: I emphasize similarities rather than differences between humans and other animals. What works in treating people who suffer from persistent anxiety? What serves to lessen obsessive-compulsive tendencies in humans? Could the same treatments work on pets?

I call my approach One Medicine.

Using the principles of One Medicine, I've treated anxiety in pets with human antianxiety medication, obsessive-compulsive disorder with prescription anti-obsessional drugs, hyperactivity with stimulants, complex partial seizures with conventional anticonvulsants or an anticonvulsant herbal derivative. If I were a physician, none of these measures would be noteworthy. But I am a professor of veterinary science, and I've experienced pushback from my fellow scientists, including reactions from raised eyebrows to out-and-out obstruction.

Once I was on a conference call with other scientists to decide funding for research into OCD, obsessive-compulsive disorder. I ventured that a proposed genetic study employing dogs as a model of the human condition was worthy of funding. A lively discussion ensued among the assembled throng and I fielded several questions through the speakerphone. There I was, Professor Nicholas H. Dodman, BVMS, DACVB, a veterinarian wired in to a large gathering of MDs—most of them psychiatrists specializing in OCD treatment.

Then came a disembodied voice from one of the psychiatrists on the line. "You will never be able to discover anything about human beings by studying dogs," the voice said sternly. "I am going to blackball this grant." I never nailed down who the speaker was, but to me, he represents a whole host of pigheaded conservatives who remain mired in the past. His voice haunts me. I am determined to prove him wrong.

On a kinder and gentler note, I recall a long conversation I once had about animal intelligence with a former Harvard professor named Dr. Marc Hauser. Trim, lively, and goateed, Marc was deep into the study of primate behavior and animal cognition. We agreed on practically every aspect of animal intelligence, including an animal's ability to learn in various ways, to modify its behavior according to circumstance, and its experience of primary emotions.

A final step, however, that Marc was not prepared to concede, was to accept that an animal may have what are termed "secondary emotions," reactions such as shame, depression, or dismay, which arise after an initial blast of primary emotions such as fear, anger, or sadness. Secondary emotions are widely considered a human-only realm. They stem from more complex chains of thinking, are oftentimes mixed, and seem to require sophisticated mental processing.

Professor Hauser readily accepted that secondary emotions *might* occur in animals. But his bet was that they did not.

"When we see a starving child on the television," Marc said to me, "we feel compelled to do something to help. That's empathy, which promotes altruism. Show me a dog who sees another dog in distress and appears upset for that other dog and acts distressed or tries to help and that may change my mind."

A couple of years later, of course, a celebrated video appeared on YouTube showing a dog dragging an injured second dog by the scruff to safety from the fast lane of a three-lane highway. Additionally, there's documentation of a mother cat going back into a burning building to rescue her kittens. A book about Ginny, *The Dog Who Rescues Cats*, tells the story of a schnauzer–Siberian husky mix who saved more than

900 sick or injured felines. There are many, many examples of animals behaving altruistically. You likely know some stories yourself.

So what in the world is the problem here? Why the resistance? At times it seems as though some scientists are the ones with conditioned reflexes, not animals. They hold stubbornly to their mistaken beliefs, all evidence to the contrary.

The answer goes back to the views of the philosopher René Descartes in the seventeenth century. Descartes avowed that animals are automatons, incapable of cognition or emotion. Such a view led to horrific cruelty to animals because the scientists told themselves that the creatures could not feel. They actually interpreted the awful screams and bellowing during vivisection and other inhumane practices as mere meaningless reflexes.

The views of Descartes were later reinforced by C. Lloyd Morgan, an extravagantly bearded grandee of British zoology, a professor of psychology and ethics at the University of Bristol in the early twentieth century. He developed a fundamental rule that he applied to the study of animal behavior. In essence, his idea was a variation of the famed "simple is best" principle called Occam's razor. It became codified under the lofty name of Morgan's Canon:

> *In no case is an animal activity to be interpreted in terms of higher psychological processes if it can be fairly interpreted in terms of processes which stand lower in the scale of psychological evolution and development.*

In other words, don't look for human parallels in animal behavior. Morgan's Canon later became an unalterable law, as though it had come down from Mount Sinai inscribed on stone tablets. The "default" position in animal research became the automaton model: no secondary emotions, no cognition, no questioning of the canon.

Skepticism has an honored and well-deserved place in science. Morgan's Canon turns a skeptical eye on all those who would anthropomorphize animals. Perhaps it is proper to wince when Mrs. Magillacuddy tells us that her cat speaks to her with its eyes. A similar principle in literature warns against the "pathetic fallacy": attributing human emotions to inanimate things ("the sky scowled," "stubborn was the looming cliff"). Sloppy thought and sentimental conclusions are surely elements to be guarded against.

But Morgan's Canon can no longer exist as unchallengeable law. The time for such a limited view of animals is long past. Instead of keeping scientists on the straight and narrow in the quest for truth, it has now become an inhibitory maxim. Such restrictive doctrines not only curtail promising humane research but also directly contribute to the wholesale killing of countless animals.

I fight the good fight every day in my job as director of the Animal Behavior Clinic at the Cummings School. Our work has one main purpose: to create happier, better-behaved animals so that owners are less inclined to give up on them—and give them up. Our efforts focus on the striking similarities between animals and human beings, both in their behavior and in their emotional and psychological problems.

We use what we discover in our research and clinical work to design treatment programs for the animals that alleviate a wide range of conditions, many of them instantly recognizable from their prevalence in humans. These include aggression, fears and phobias, obsessive-compulsive behaviors, canine autism, depression, post-traumatic stress disorder (PTSD), and Tourette's syndrome.

On a regular basis, I diagnose psychological conditions in my non-human patients that also occur in humans. These behavioral issues might be readily treatable, just as they are in human medicine, but most animal owners are simply unaware of the possible options. Once the diagnosis is clear, the treatments can be simple and straightforward.

Sometimes behavior modification training is the answer or sometimes small changes in the pet's diet or environment. Sometimes a prescription drug is necessary. Many of the drugs we use with animals are the same as those used by medical doctors for human patients.

If we want to prevent unnecessary relinquishment of pets to shelters and pounds—as surely we all do—then word must get out about these successful treatments for unwanted, troublesome behaviors. That is why in this book I'll tell you about some of the most interesting emotional problems that I have encountered in a variety of species.

The word for my profession came into the language in the seventeenth century, from the Latin *veterinarius*, which means "of or having to do with beasts of burden," but in colloquial usage meant "cattle doctor." Ordinary creatures such as dogs and cats were easily dismissible in the 1600s, but cattle, well, cattle were cash cows, so to speak. People depended on them for their livelihood, and so early on they were deemed worthy of doctoring.

But because the veterinary profession has generally focused on doctoring, they haven't paid much attention to treating—or analyzing or understanding—unwanted behavior in pets. Behavioral medicine still goes untaught by the majority of veterinary schools and its core principles remain a mystery to many veterinary practitioners. The American Veterinary Medical Association took until 1993 to establish the American College of Veterinary Behaviorists. Of the thirty or so vet schools in North America, only a dozen or so regularly teach their students about clinical animal behavior. A fledgling vet can easily graduate without significant training in this vital area—and that's a great shame.

I was shocked to hear a professor of veterinary medicine say that "animals are automatons and their behavior is simply a result of a series of conditioned reflexes." He was a cardiologist, but may as well have been a car mechanic for all the insight and empathy he displayed. Another veterinarian once said—in public—"You don't have to like animals to be a veterinarian, just as you don't have to love toilets to be a plumber."

If I ever heard such sentiments from a vet about to treat my own pet I'd run the other way. I believe without a doubt that empathy and affection for animals go a long way toward making a good veterinarian.

The practice of One Medicine preserves scientific rigor, but at the same time embraces the obvious shared biology of human beings and nonhuman animals. Consider Morgan's Canon not as an unalterable law, but as a human-created principle that can be useful and limiting in equal measure. Open our minds, open our hearts to the unconstrained possibilities in the interplay of pets and people.

Can we treat animal behavior with the same medicines, therapies, and approaches we use on humans? To me, the answer is an obvious yes. You could effectively teach medical students brain anatomy using the brains of dogs. Transferring what you learned about dog brains to knowledge of human brain anatomy would be a breeze.

It's not just brain anatomy that is similar across the species. Animals in general, and mammals in particular, have many of the same inner workings as humans do. We share physical similarities and we process and respond to incoming sensory information in much the same ways. Under the hood, so to speak, in terms of the nervous system or other organ systems, there is not much difference in how things work. I'll tell you more about how we can learn about animal behavior by carefully examining the animal who stares back at us from the mirror.

Why use a pill developed for humans to treat behavior in animals? It may seem outlandish, yet all our medicines—*every single pill humans place into their mouths, every injection, every balm, every suppository, even*—was first tested, developed, and refined using animals. Are you a man or a mouse? Well, in pharmacological research, mice lead the way, and humans follow.

It's happening right now. As you will see in the following pages, One Medicine has already yielded astonishing results. We stand on the cusp of a revolution. The excitement and optimism I feel are intoxicating. I hope they turn out to be contagious.

CHAPTER TWO

The Horse Who Chewed on Fences

Obsessive-Compulsive Disorder

The question is not, "Can they reason?" nor "Can they talk?" but "Can they suffer?"

—JEREMY BENTHAM

Let's bring in the proverbial Martian to observe human behavior. And let's suppose that communication with such an interplanetary being proves impossible, because our languages don't mesh. How will the Martian possibly understand us, when all it has to go on is what it sees with its own three eyes?

Our Martian comes around a corner and suddenly confronts a man pounding his head against a brick wall. By observing this behavior, the thought might arise in the Martian's mind, *Something is wrong here.* Such a conclusion arises naturally, and no common language is necessary.

We folks who study animal behavior are in much the same boat as our space alien visitor. Veterinarians share no common verbal way to communicate with animals, because our languages don't mesh. In order to understand what's going on, we humans in the white lab coats are limited to our observational abilities. Yet determining things about animals without verbal communication is what we veterinarians do best. The verb "to vet" derives from "veterinarian." To vet means to examine thoroughly and extrapolate from what we see. That's always how we doctors seek to function. And just like the Martian, we recognize certain emotional states in animal subjects.

Yet, if we're scientists, we must not say no. Here is the voice of traditional science: *If it can't be measured, it isn't real.* If you want to be a "real" scientist, you can't ascribe any quality to animals that you can't measure. Emotions are the realm of humans. Remember Morgan's Canon? I am not allowed to suggest a cocker spaniel feels sadness when its owner dies. Or that a cat demonstrates happiness when she greets her human companion. No, no, no. We are supposedly looking at the mere mechanical workings of wind-up toys.

Some horses indulge in the strange repetitive behavior called "cribbing." You can find disturbing videos of it on the Internet. Whenever and however you encounter cribbing, it's the equivalent of the Martian confronting the human headbanger. Anyone with intelligence concludes instantly that something is seriously wrong.

I knew about cribbing, but I still reacted strongly the first time I saw it. An otherwise healthy-looking palomino mare with the fabulous name of Poker's Queen Bee, was positively addicted to biting her stall.

Every few seconds Poker would lean forward and anchor her teeth in the edge of the stall door. Her incisors dug powerfully into the wood. Then she would lean backward and swallow hard, making a loud grunting sound—a behavior called "wind-sucking"—as she pulled with all her might on the stall.

The pathetic fallacy would have us believe that Poker was crying out against her captivity. It was as if she was trying to pull apart her stall. *Let me outta here!*

Cribbing is indeed a condition of confinement, the product of an unnatural lifestyle with fettered natural outlets. More specifically, though, the issue lies in the eating department. In the wild, horses never crib—and not just because they don't live in stalls. Free-ranging horses graze for 60–70 percent of their day. Cribbing represents a frustrated response that becomes ingrained. Horses that crib are reacting to a lack of an opportunity to graze on grass.

But we don't have to know all this to understand Poker's behavior on a more basic level. Even a child (or a Martian) would comprehend the situation after witnessing her bite, lean, and grunt. *Something is wrong here.*

Poker's owner, a Boston-based business executive, was concerned about Poker's health. The practice wears down teeth and can lead to digestive disturbances, such as life-threatening colic. Understandably, she also found the behavior annoying.

Vets with little feel for animal behavior have developed various vile treatments for cribbing. They've implanted brass rings in the offending animal's gums, or had the horse anesthetized in order to surgically sever its neck muscles. They've even tried boring holes in its cheeks. These painful, worse-than-useless techniques stem from the automaton model of animal behavior. It is as if those vets were considering a mechanical toy that was broken. Surely simply bending this spring there or adjusting that ratchet there will do the trick.

Step back a minute and consult your common sense. Brass implants? Sliced muscles? Drilled cheeks? Common sense responds: Are you kidding me? These measures are not only ineffective. They're inhumane. It's as if the Martian were to strap a brass plate onto the head of the poor guy bashing his head against the wall. Um, no, that won't do.

Owners of cribbing horses become frustrated enough to try almost anything. Some take the behavior personally, concluding their horses are just being brats, cribbing simply in order to spite them. Nothing could be further from the truth. But cribbing does have far-reaching physical consequences. Horses that indulge in it fail to thrive and cribbing decreases the animal's value in the marketplace. This alone can have perilous consequences for a horse. Off to the slaughterhouse or glue factory!

Cribbing was the first behavioral condition I investigated in my career, even before my study of nighttime aggression in dogs like Comet. At that time I was that proverbial Martian, limited in language but clearly grasping the fundamental truth of an animal in distress. Trying to understand the condition, and ultimately seeking to lessen the numbers of horses sold, neglected, or euthanized because of it, my Tufts University colleague, Dr. Louis Shuster, and I embarked upon a study of Poker and other obligate cribbers.

It was early summer in 1984. Lou was then a senior professor of pharmacology at Tufts Medical School. Fifteen years my senior, he had an impressive résumé of work done on the topic of drugs of addiction. His professional standing was recognized by the National Institutes of Health (NIH), which enlisted Lou to help select the proposed research projects worthy of funding. As a scientist, he was head and shoulders above many of his colleagues, including me! I was lucky to have him as a friend and mentor.

A veterinarian paired with a medical professor-researcher: To some, we made for an odd couple: humans is humans and critters is critters and never the twain shall meet. But Lou grasped the principles behind One Medicine and demonstrated a quality that is supposedly a prerequisite for scientific work, but crops up all too rarely in labs and research facilities: he possessed an open mind.

At our invitation, Poker's owner trailered the horse up to the Cummings School at Tufts in North Grafton, Massachusetts. We brought the mare into the equine ward of our Large Animal Hospital, a bright, cheery place with the sweet smell of hay battling it out with the unmistakable odor of manure. Poker negotiated the rubber-mat-surfaced walkway to her fifteen-by-twenty-foot stall, which featured vertical metal bars on a wood-paneled door. After she explored her stall, she located a wooden crossbar and began—predictably—to crib. The dismal sight of the horse's obsession at work also triggered a reflexive response in me, one of dismay and urgency. We had to help this poor animal.

Lou and I were aware of being in the presence of a truly magnificent beast, though one who was clearly disturbed. I stroked her neck and did my best to calm her, but she continued to crib. We had to embark upon our investigations with a subject who was in the viselike grip of the very symptoms we sought to alleviate.

Our first step was perhaps the most invasive. To prepare Poker for treatment, we needed to insert an intravenous catheter into her jugular vein. This wasn't a difficult procedure, since a horse's jugular is large and easy to locate. (Mongol warriors, who lived more on the saddle than on the ground, would lean over and prick a vein of their mounts, in order to be able to drink the nourishing blood without having to halt.)

The neck catheter was necessary, because we would inject medication through it that we hoped would stop the cribbing. I prepped the area with antiseptic and a topical anesthetic, and then inserted the catheter into the vein. Poker, a stoic by nature, did not flinch during the procedure, barely noticing the incursion. I then capped the device and flushed it with saline. We were ready to go. The white plastic catheter nesting in the furrow of Poker's neck was an odd sight.

Before medicating Poker, we began a count using a mechanical counter to see how often she cribbed. The horse was relentless. Her rate of crib biting was around 250–300 times *per hour*. That averaged out to four or five times a minute for the period we studied her. She religiously observed her holy trinity of bite, lean, and grunt. Poker didn't take too many breaks from cribbing, though occasionally she would saunter around the stall, grab a few mouthfuls of hay off the ground, and then peer out of the stall door, apparently taking in the surrounding environment. But then her obsession would take hold once more. When we plotted her activity on a graph, it showed a straight line upward with no pauses. She barely stopped, even to eat.

Poor Poker. It was distressing, to say the least. Clearly, the animal was in extremis. And we were desperate to make her feel better.

Our first "treatment" was the medical equivalent of a feint. We injected Poker with twenty milliliters of saline, a placebo, to make sure the process of injecting her didn't itself cause a change in her behavior. The injection made no difference. On she cribbed. The chewing, lunging, and wind-sucking made a constant din.

Next we injected Poker with naloxone, a morphine antagonist drug—that is, a medication that works to cancel out the effects of morphine. It took a few minutes, but her cribbing stopped completely.

The sudden silence after the continuous sound of cribbing seemed almost surreal. It was as if a car alarm had quit after a too-long stretch of loud, annoying noise. Standing there in the sudden silence of the barn at Tufts, Poker's owner shed tears of joy and relief. We were all moved that the horse's distress had abated so swiftly.

Lou and I had done it. We had successfully halted a harmful behavior in animals with the administration of a medicine, naloxone, developed for use in humans. But of all the meds in all pharmacies in the world, how had we arrived at that one? The answer involves some explanation.

Cribbing is a reaction to stress and the cribbing behavior— while distressing to people watching—actually stimulates the release of endorphins, or soothing, pleasurable brain chemicals. The horse feels better doing it—even though the habit, like many habits, is not good for the horse. The endorphin release reinforces the behavior, so the horse keeps doing it. The horse is, in effect, "self-medicating" to cope with the stress of confinement. The abnormal behavior becomes habitual because the horse is hooked on her own natural, internal morphine—those endorphins.

Naloxone—trade name Narcan—which we used on Poker, is in the news of late as a treatment for drug overdoses. Public health advocates have urged equipping police and emergency medical teams with the drug so that first responders could save thousands of lives. Such sensible measures often run up against crude prejudices: Why should public money be used to rescue drug addicts? But lately reason and logic have pretty much won out. Save lives first. Deal with politics of addiction later.

The idea to treat Poker with naloxone arose from a previous study I'd done with Lou, shortly after I joined Tufts, on "reverse tolerance." Some animals (most of Lou's experience was with mice) become increasingly sensitive to morphine's effects after repeated injections. In other words, they get more affected with every subsequent dose. As with mice, so with horses. Other researchers unwittingly documented the same reverse-tolerance effect in a study performed on Kentucky thoroughbreds.

Reverse tolerance is not well-known in humans. In fact, addicts are commonly understood to develop the opposite effect, acquiring a *tolerance* to morphine. They become desensitized, meaning they need larger doses to produce the same effect. But that wasn't what was happening with Lou's mice, who became increasingly sensitive to morphine, not less. But it turns out that reverse tolerance to various substances does indeed occur in people. After a "drug holiday," or medication vacation, for instance, amphetamine users can feel increased stimulation (in addict lingo, a "bump") and a decrease in side effects.

Once Lou and I had figured out how to help Poker, we were able to rescue some horses from a slaughterhouse in Connecticut in order to expand our investigations into endorphins. These horses, too, experienced bumps of their own when given serial doses of morphine: they circled or paced around the edge of the stall, dug at the ground, and mouthed the surfaces in their stall—engaging in cribbing behavior and other stall vices, as equestrians call them.

So if increased sensitivity to morphine triggered these behaviors, then stall vices were likely fueled by nature's own morphine-like chemicals, endorphins.

Now we needed a way to block the action of endorphins, which is what led us to try naloxone on Poker. Morphine antagonists oppose, interfere with, or halt the effects of morphine and of naturally occurring endorphins. If we gave a cribber a drug that acted as an antagonist, perhaps the medicine would stop the behavior in its tracks. And it worked.

What happened with Poker's Queen Bee also occurred with many other horses in subsequent trials. The cribbing completely shut down after an injection. For extended periods of time, we could mark "zero cribbing" on our charts.

This first study on cribbing began the kind of inquiry that to this day I still conduct in my life as a veterinarian—formulating a hypothesis, testing it in our trial at Tufts, then reporting the results professionally. All of my studies have been based on a fundamental insight into animal behavior, the profound recognition that we animals share the same neurochemistry. Our minds work in similar ways.

When we started working with Poker, none of us realized that our experience would change all our lives, transform our professional focus, and have far-reaching consequences for the study of animal behavior. Poker Queen Bee's owner quit her fancy job and went back to Loyola University in Chicago, where she earned a PhD in biochemistry. Dr. Shuster, who had an outstanding career researching addiction to drugs changed his focus to study addiction to bad behaviors.

As for me, well, I had found my true calling: a lifelong study of animal behavior and the pursuit of One Medicine.

———————

Science is sometimes like love, in that the course of research never does run smooth. In the course of our work with Poker's Queen Bee, we were able to reduce cribbing to zero. Success! Unfortunately, the compulsive actions started up again when the drug wore off. Setback!

With One Medicine we can easily see significant similarities in conditions shared by animals and humans. Obsessive-compulsive disorder, or OCD, for instance, can be considered a condition of degree: that is, normal behavior amped up to an abnormal degree. Human OCD arises out of deep-seated behaviors that are necessary for survival, but are taken to an extreme.

Humans are a hunter-gatherer species. Hunting was necessary for the survival of early hominids. It required some risk taking, but also caution, so that the hunter wasn't exposed to extreme physical danger. Gathering was also an essential activity for our ancestors. These behaviors are fundamental, and when successful were passed along to successive generations, until it evolved to be programmed into the human brain. These activities are normally employed appropriately when the need arises.

But when the behaviors are triggered continuously and out of context, an obsessive-compulsive disorder results. Since staying clean conferred life-and-death survival benefits, washing became a normal, customary behavior, hardwired into us, but people often manifest OCDs as undue fixations about personal cleanliness, with excessive hand washing and showering as well as compulsive skin picking and nail biting. Excess caution can lead to a preoccupation with germs and cleanliness. Too much of a good thing, so to speak. In the same manner, traditional food-storing behaviors transform into the aberrant practice of hoarding. All OCDs reflect excessive expression of basic, species-typical, survival-related behaviors.

Human society has known about OCD for a very long time. Fortunately, the days when such behaviors were ascribed to the work of the devil are over. Nowadays, the condition is considered to be a spectrum of disorders. Human OCD is no longer thought of as linked only to concerns about germs, personal safety, or possessiveness. Other compulsions are included under the same diagnostic tent. The tent is a virtual big top: compulsive shopping, kleptomania, body dysmorphic disorder, bulimia nervosa, anorexia nervosa, obsessive checking, touching, ordering, and arranging, trichotillomania (hair pulling), pyromania (fire-setting), and various paraphilias (extreme sexual behaviors) are now considered by some to be related in their underlying mechanism.

It turns out that these behaviors are based in specific areas of the brain: the limbic system (center of emotions) and basal ganglia (a repository of species-typical, hardwired behaviors). Because of the similarities between human and animal brain function, it's no surprise that over time we found numerous compulsive disorders in pets. Besides cribbing in horses, such disorders can include compulsive licking and tail chasing in dogs and cats who strip out their hair. At first, the veterinary profession didn't connect these extreme behaviors to OCD.

A huge gray horse named Mobey, formerly on the Bermuda show circuit, had been retired for constantly bucking his owner. The animal found his way to Tufts, where he was housed in one of our equine wards as a "teaching horse." Mobey was sincerely loved and admired, and our students performed many, many practice examinations on him. He went a long way toward proving the old adage that an unexamined life is not worth living. Eventually, however, the money ran out to keep him, and he became a candidate for euthanasia.

A single quirk saved Mobey's life and pulled him back from the edge. He was a confirmed cribber, constantly biting the wooden boards at the front of his stall. If left to his own devices he might have dismantled the whole barn. All we could hear, as we entered the ward, was a great rattling of the stall, followed by the familiar grunting sound as Mobey gulped air.

Lou Shuster and I opted to pay for his keep. We wanted to replicate with Mobey the treatment we had tried with Poker's Queen Bee. More than that, we aimed for a longer-lasting remedy, one that would extend beyond the temporary fix we had given Poker.

We began our effort by using different morphine antagonists to stop Mobey's cribbing. Even though we anticipated the same results, they were still remarkable. Each drug worked. After we'd tried a series of different drugs, we arranged for the meds to be delivered by constant infusion using a portable pump strapped to his neck. As long as the infusion remained above a certain level, Mobey did not crib. When we turned the infusion rate down below that level, Mobey would start cribbing again. It was as if we had a radio-controlled horse. Turn the knob to the right and he stopped cribbing. Turn it to the left and he would start up again.

But the solution had its own set of problems. One of the most pressing was that the treatment was not practical, because opioid antagonist drugs are expensive. They cannot be given by mouth because they are rapidly destroyed by the liver. And the effects of a single injection—as opposed to Mobey's pump infusions—lasted only a short time. Around four hours of cribbing suppression was about the best we could achieve when we administered the injection.

To try to resolve this problem of duration, Lou and I experimented with various long-acting preparations. Taking a cue from the wider world of illicit drug use in humans, we found ourselves applying a "freebase" of the drug directly to mucous membranes inside Mobey's mouth or nose. Using a long-acting injection of one of the morphine antagonists, we finally did manage to eke out almost a month with no cribbing. The prohibitive cost, coupled with a lack of interest from suitable sponsors, caused our project to end.

The good news is that Mobey was eventually adopted out of Tufts to a local stable. We dodged the need for him to be euthanized, and he lived out the rest of his days in a peaceful environment.

While our early research in cribbing failed to yield an effective long-term treatment for the behavior, it ultimately supported the concept of One Medicine; the medical commonalities between humans and animals. Heartened, I resolved to move forward and investigate problems further. I didn't realize back then that my research would prove highly controversial. I didn't fully understand that a certain sector of the human populace would display an aggressive, knee-jerk aversion to the ideas of One Medicine. Scientists and laypeople alike preferred not to be reminded that they were animals, thank you very much, and some were aghast at the news that they shared crucial elements of their physiology with their pets.

I learned my lesson in a memorable and sometimes quite public way.

CHAPTER THREE

The Dog Who Came Back from the War

Post-Traumatic Stress Disorder

> Besides love and sympathy, animals exhibit other
> qualities connected with the social instincts which in
> us would be called moral.
>
> **—CHARLES DARWIN**

One pleasant morning in the summer of 1993, Bill Doust put his dog, Elsa, in the car to get ready for a ride to the local park. A large mixed-breed pooch of around seventy pounds with a semi-long-haired, blackish-brown coat, Elsa's most marked characteristic was her sad-looking eyes. Bill parked in the shade so that the interior would not overheat, and also rolled the windows partially down to make sure Elsa would stay cool.

A construction supervisor by trade, Bill loved all animals. To Bill, Elsa was not just a dog, but a cherished member of the Doust family. The love flowed both ways. Elsa lived for him. He practically lived for her. This kind of connection of love and loyalty can rank among the deepest emotions of human life.

Several years previously, Bill had rescued one-year-old Elsa from an abusive situation. Since that time, Elsa shadowed Bill whenever he was at home, following him from room to room. She exhibited a touch of separation anxiety whenever he was away, but the Doust household provided a warm, charming environment for all creatures within its embrace. Elsa contributed a large part to making it that way.

With the dog left behind in the car, Bill ducked back inside his house for a quick shave before driving to the park. Elsa liked waiting in the car. She understood that it meant something good was in the offing.

Standing at the bathroom mirror, his face lathered with shaving cream, Bill heard a fracas outside. There were shouts, the pounding of running feet. He went to the window. Some sort of police incident was going down. A man ran past Bill's house. There was a clamor of "Stop, thief!" A squad car, sirens on high, wailed up to the scene.

Bill watched the action unfold. The suspect dashed down the block. A cop in uniform bolted from the cruiser and headed off in hot pursuit. As the officer sprinted past Bill's car, Elsa jumped out of the partly open window and ran after him.

Suspect, cop, and dog tore down the block. Bill couldn't believe his eyes. He himself raced out of his house and joined the chase. Suspect, cop, dog, dog owner.

Elsa closed in on the policeman. Fearing an attack, the cop kicked her away. The normally peaceful dog lunged back and locked her jaws onto the officer's pant cuff, snarling and worrying her head from side to side.

Bill made it down the street and approached. "Don't worry," he shouted. "She's my dog! I'll get her!"

But the policeman drew his gun. As Bill screamed, "No! No!" the cop took aim and shot Elsa in the head.

The officer pounded off, continuing the chase, leaving Elsa's inert, bleeding body on the sidewalk. Bill scooped her into his arms. He had seen the bullet strike her and was convinced his beloved pet was dead.

Even in his panic, Bill had the presence of mind to transport Elsa to a top-of-the-line emergency veterinary clinic. Working quickly, in triage mode, the clinic staff administered multiple blood transfusions, saving her life. Elsa got patched up. The officer's bullet, from a .38 caliber handgun, had entered her skull just in front of her ear, permanently deafening her on that side. The bullet continued on its way down Elsa's neck and lodged in the wall of her chest cavity. We still have the X-rays on file here at Tufts. They show a serious injury. Elsa had cheated death by centimeters.

Thanks to excellent and timely critical care, Elsa survived and recovered. But unbeknownst to Bill and the veterinarians involved, the psychological trauma of the shooting had changed her profoundly. Elsa returned home after her brush with death. She became more attached to Bill than ever, a "Velcro dog," sticking to him like the well-known fastening material. During the day, this did not pose that much of a problem, because Bill took her along with him to work. Elsa would hover around the job site, scrutinizing his every move, until he summoned her with a wave of his hand.

At night Elsa's separation anxiety became almost impossible to tolerate. Bill could not go to sleep without Elsa pawing at the bed for attention. This was a new behavior, which led to insomnia for both parties. Luckily for Bill, his grown son lived with him. The two of them adopted a tag team approach. Bill would sleep for three hours, his son took a three-hour shift, and together men and beast would somehow make it through the night.

In the aftermath of her injury, Elsa had become hypervigilant, never settling, always on the lookout for trouble. Any element of the shooting incident triggered a terrified reaction. She became a nervous wreck whenever she saw or heard a police car, heard a siren of any kind, saw any flashing lights, or met people in uniform. The police officer happened to have been African American, and Elsa now responded with fear when she encountered black men.

Bill Doust took action and sought help for Elsa, eventually coming to me at Tufts. At the time, I dubbed her condition "PTSD-like." Certainly it was posttraumatic. Elsa's nervousness, avoidance of anything reminiscent of the original psychological and physical trauma, plus her nocturnal anxiety—could that stem from bad dreams?—smacked strongly of post-traumatic stress disorder.

I treated Elsa with an antidepressant, supplemented with nighttime Valium and a nonaddicting morphine-type painkiller, butorphanol. With a few ups and downs, Elsa improved. But she would never completely get over that dreadful day. Fears once learned are never forgotten, although it is possible to ease them by superimposing new learning, with or without the help of medication.

Over the ensuing years, I saw many other PTSD-like cases in dogs. In my behavior course, I began to teach PTSD to Tufts veterinary students as a bona fide canine condition. I didn't realize I was violating the sacred principle of human exceptionalism. I simply saw a situation and attempted to address it with reason, compassion, and education.

Then came Gina.

A two-year-old German shepherd, Gina had spent six months on tour in Iraq, accompanying a unit of the US Army. Her duties there focused on searching buildings for explosives, working with troops who were "cleaning out" insurgents from their hiding places, always performing in chaotic and complex urban environments. Before entering a previously unsearched structure, Gina's human comrades would often toss flash grenades—nonlethal charges that explode in a flash of light and a high-decibel report. Once while she was riding in a convoy, an IED, an "improvised explosive device" or roadside bomb, exploded nearby.

When Gina returned from her tour, she was fearful, disturbed, and psychologically fragile. The handler to whom she was assigned back in the States, Master Sergeant Eric Haynes, the kennel master at Peterson Air Force Base in Colorado Springs, gave an interview in which he described her condition as post-traumatic stress disorder. An army veterinarian later concurred.

The press went wild. How could a dog get PTSD? Surely PTSD was a psychiatric diagnosis reserved strictly for humans?

Here is testimony from an expert interviewed by the media at that time:

> *A psychologist on the faculty at Columbia University's Mailman School of Public Health, said PTSD is a diagnosis developed for humans, not dogs. "That's not to say that animals can't be traumatized. It sounds like this dog was traumatized from the experience of extreme stress and fear," he said. "That causes an alteration in the animal's nervous system similar to an alteration of the nervous system in humans."*

The Columbia guy got it right about changes in the nervous system in response to trauma, but he could not make the logical leap and call the condition by its proper name.

The press, when it covers a breaking story such as this one, becomes something like a bloodhound. Rabid for additional sources, reporters sought out anyone who would comment on the outlandish idea that a dog could suffer from a condition heretofore reserved for humans. They turned up my name, and I agreed to an interview.

Yes, I had seen it before, I told the journalist who called. Yes, PTSD can and does occur in dogs. Yes, indeed, I teach about canine PTSD in my course to veterinary students. I would consider myself remiss if I didn't.

While I did not examine Gina firsthand, I speculated that her experiences in Iraq were the type that could lead to post-traumatic stress. Before going to war she was a happy, playful, confident dog who loved her work. Training sessions all seemed to her like great fun. Then she was deployed to Iraq and she faced extreme, violent, and unsettling real-life situations. This was a whole different ball game. Her training could not have prepared her for it. With Gina at their side, the soldiers she was with did what they had to do, kicking down doors and throwing grenades to shock and confuse the enemy.

Check out this description of such an explosive device:

> *Upon detonation, stun grenades emit an intensely loud "bang" of 170–180 decibels and a blinding flash, sufficient to cause immediate flash blindness, deafness, tinnitus, and inner ear disturbance.*

Troops wear goggles and ear protection to shield them from the extreme visual and auditory assault meted out by the grenades. War dogs are not similarly well-insulated from such severe sensory overload. Even beyond the sound of explosions, a war environment is one of constant sensory assault. Yelling, door-kicking, and chopper noise form a brutal soundtrack to the missions, and exhaust fumes and other unfamiliar smells abound.

No wonder Gina became unnerved. Anxiety overwhelmed her. She became terrified of people. After returning from Iraq, she could no longer perform her duties during training sessions. She balked at entering homes when directed, stiffening her legs and pulling back at the threshold. When forced to go inside, she would slink and cower, often immediately taking shelter under the nearest piece of furniture.

In the wake of trauma, Gina also became hypervigilant, always on the lookout for trouble. Her handlers said she suffered the "full range" of PTSD symptoms, which would also include sleep disturbances like Elsa. Gina had developed what Sigmund Freud labeled "war neurosis," which later became known as "shell shock."

Like Elsa, Gina recovered after spending some months of treatment in a safe environment. She went back into training and managed to handle it without balking. The US Army is a stern taskmaster in the normal course of events, and often sends soldiers suffering from PTSD back into war zones. The army had great hopes for Gina going back into service, but I doubted that she would ever be satisfactorily rehabilitated. The memories of what went on before would forever haunt her. Gina's condition might not count as a reason for honorable discharge, but I think that it should.

Let's consider the human condition of PTSD. According to the DSM,* PTSD occurs after a person has been exposed to a traumatic event, such as witnessing or experiencing events that threaten death or serious injury to self or others. To be classified as PTSD, the response of the victim must involve intense fear, helplessness, or horror, which sounds similar to what Elsa and Gina went through. People with PTSD also have dreams and flashbacks, which we cannot verify in a nonverbal species, but recurrent dreams of the event may explain nocturnal agitation in dogs. Clearly apparent for both Elsa and Gina were symptoms straight out of the DSM, including "intense psychological and physiological distress on exposure to external cues that symbolize or resemble the traumatic event." They also avoided stimuli associated with the trauma, another basis for the diagnosis of PTSD in people, as well as difficulty falling or staying asleep, difficulty concentrating, and hypervigilance. Finally, they seemed detached and disinterested in life after their trauma.

To qualify for a PTSD diagnosis, the duration of the disturbance must be greater than one month. During this period, the condition

---

* The *Diagnostic and Statistical Manual* (*DSM*), an oft-revised and updated standard that classifies mental disorders, is used by psychiatrists, clinicians, researchers, and other health professionals.

must cause clinically significant (that is, measurable) distress or impairment in interactions with others. Gina and Elsa both qualified in duration and degree of distress. On almost all counts, dogs with PTSD display syndromes comparable to those of affected people. And we can establish this even without the benefit of actually speaking with the subject.

Although Gina was a dog of war with a classic history of "battle fatigue," Elsa had never been to war. People don't need to go war to develop PTSD, either. Any traumatic events can trigger the condition. Street violence and other forms of "violent personal assault" (such as Elsa experienced) are also well-established causes of PTSD. Incarceration as a prisoner of war, confinement in a jail or concentration camp, abusive relationships in childhood and adulthood—all can trigger the disorder. We don't have to search too far to find a parallel in dogs and other animals. Brood bitches in puppy mills, for instance, suffer PTSD as a result of their long isolation in pens and forced breeding.

Automobile accidents can cause PTSD in both people and pets. A serious accident of any kind necessitates treatment in an intensive care unit, which itself could create unpleasant memories. A stay in the ICU, especially one that involves distressing experiences, such as being intubated, paralyzed, and ventilated, can trigger or compound PTSD in people and in pets.

At Tufts we treated a dog, Star, who had developed an extreme panic reaction to visiting veterinary hospitals following a painful treatment for a grievous injury. The upshot: Star wouldn't even get out of the car a hundred yards from the entrance to the vet hospital. The location itself triggered her distress. The dog's regular vet made house calls for a while, to avoid causing the dog the anxiety of office visits, but the owner wanted to cure Star of her fear of clinics.

To me, it was clear that Star had PTSD and my approach was similar to the human treatment for PTSD. I prescribed a short-acting, as-needed medication, clonidine, to assuage Star's strong fear reaction. And I also suggested a long series of desensitizing retraining actions, in which the owner would gently reintroduce the dog to the experience of hospital care in stages. She should not force Star into the situation, as this would only aggravate her condition. The dog's owner dutifully took on the task, emailing us every week with the latest news. Progress occurred gradually over the course of a year, and Star will now walk into the vet's office, tail wagging, and will take treats from the staff.

PTSD in people can be triggered by seemingly harmless situations that are reminiscent of the initial trauma. I've encountered a similar trigger for canine PTSD. For instance, the smell of cooking lamb triggers terror in some dogs. Dogs that are frightened by the scent of lamb chops on the stove may be reliving some traumatic event that occurred. Perhaps they burned their noses while investigating the source of the formerly delectable odor. One trainer amusingly suggested that the condition may be restricted to sheep dogs distressed by the smell of the sheep they protect and herd being cooked.

Not every person or every dog exposed to psychological or physical trauma will develop PTSD. Some soldiers and some military dogs return unscathed after multiple tours of duty. Others seem primed to acquire this debilitating psychological condition. The reason for this may be something to do with genetic predisposition, which in conjunction with environmental events, may affect a person's (or a dog's) response.

Resilience has been the subject of extensive psychological testing, and researchers have determined that people with a positive outlook on life bounce back more quickly in the face of multiple life, medical, and psychological challenges. This suggests the inviting possibility that there might be a "resilience gene" lurking somewhere on the human genome. If we could find the genetic basis for resilience, we would have new avenues to treat any particular behavior causing distress by tweaking the resilience system.

As exciting as this is, much of the relevant research in lab animals on PTSD has already been accomplished. But it has simply not been taken up by the medical community as a whole. Attention must be paid.

Genes expressed in the central nervous system make proteins that activate neural pathways. If we knew what genes were involved in the response to trauma, we could predict PTSD susceptibility and explore ways of preventing—and treating—PTSD in man and animals.

For example, in traumatic situations, animals release large amounts of stress hormones and neurotransmitters. Adrenaline, the fight-or-flight neurotransmitter, enhances memory of highly emotive events. This makes evolutionary sense. When an animal acutely remembers a place or situation that is dangerous, and avoids it in the future, it is more likely to survive than a counterpart whose memory of those events is less sharp. The latter remain more likely to stumble into the same dangerous situation again. The next time could be the last time.

But in treating cases of PTSD, we seek to reverse such fear-based learning. If fight-or-flight brain chemicals enhance memory, then blocking their effects with a so-called beta-blocker, such as propranolol, should prevent acute and lasting memory acquisition.

This turns out to be the case. Propranolol prevented the imprinting of stress-induced learning in rats. Beta-blockers do the same in people. The brain region central to high-stress learning is called the amygdala. Activation of the amygdala enhances fear-induced effects, so tamping down the "vigilance" of the amygdala reduces anxiety throughout the brain.

At one time it was thought that taking a beta-blocker *before* a predicted trauma was the only way to avert stress-induced imprinting. That conclusion was drawn from studies of rodents and clearly was not convenient for a military application. But it has now been shown that treatment with beta-blockers *after the fact* can also prevent the brutal "flashback" aspects of post-traumatic stress. This finding renders beta-blocker treatment all the more compelling, as it points out a way of addressing PTSD in people and in animals.

So soldiers in battle need not take a beta-blocker three times a day, "just in case." They could take a pill within a few hours of fighting or being involved in a horrific event to ward off PTSD. Beta-blockers can prevent the experience of trauma from leaving a deep psychological wound. Prevention—even slightly after the fact—is better than a cure.

Since the recent conflicts in Iraq and Afghanistan, we in the United States find ourselves with an abundant supply of PTSD subjects: both our human veterans from the armed forces and their canine support troops. We simply need sponsors for studies of war dogs to help us better understand the condition in humans and canines. If only we could get funding to heal the traumas of war as easily as the Army is able to secure funding to make war.

It would be much more enlightening to study PTSD in war dogs than in lab rodents. Because dogs develop the condition after being in similar situations to those that traumatize soldiers, they would provide a more accurate, real-life study. We could study which dogs were more susceptible and why and explore the genetics and neurobiology of that susceptibility. And we could develop new and better treatments—for dogs and people.

Clearly, the adrenaline system is "switched on" in PTSD when an animal or person is exposed to a traumatic event. Subjects experiencing the same traumatic event can have widely different responses. One might develop PTSD while another does not. This implies that a factor in the genetic makeup of the affected subjects may render them vulnerable to the disorder. And that, in turn, points the way to a genetic-based treatment. If we can find a gene that figures into the process of PTSD, we can figure out who is vulnerable and treat them accordingly.

At Emory University in Atlanta, Georgia, a team investigating PTSD focused on just such a genetic approach to the problem. Unlike the folks involved in the kerfuffle over Gina's PTSD, the Emory research team members have no issue with animals having the disorder. And they willingly extrapolate from animals to humans. Testing on animal subjects and extending the findings to humans pretty much describes how biological research science is normally conducted, and that kind of back and forth sums up One Medicine in a nutshell.

The Emory team used mice in their study, the tried-and-true subjects that have spurred so many scientific advances. They stressed the mice by immobilizing them so they couldn't seek cover or hide, which made them afraid. Researchers then released them into a maze through which they had to find their way. Mice that had been stressed by immobilization had trouble finding their way out of the maze, indicating long-term memory impairment, which is also found in cases of human PTSD.

In order to find the pathways underlying the fear response in the amygdala, the Emory team focused on a specific gene. Going by the technical name of Oprl1, the gene influences numerous brain activities, particularly instinctive and emotional behaviors. The researchers basically showed the physical pathway that fear takes and what it does to the brain. This is what real science can do today, penetrate deeply into areas heretofore unavailable to research. We aren't limited to cells or even genes anymore. We deal in molecules.

The discovery at Emory suggests that effective treatment can be developed to heal the specific pathway of fear. The endgame, of course, is to come up with effective treatments for PTSD-afflicted service personnel, and for war dogs such as Gina and traumatized animals such as Elsa.

The Emory researchers also looked at the Oprl1 gene in humans, both PTSD and anxiety sufferers. In research studies where subjects viewed pictures of fearful faces, the amygdala lit up more brightly in PTSD-susceptible people, as determined by changes in the Oprl1. So changes in this specific gene may increase the likelihood of developing PTSD. This suggests that measures taken to limit those genetic changes might be beneficial.

The new discoveries have determined the reason why some animals and people are susceptible to developing PTSD. The Emory scientists nailed down the gene and neural pathway that promotes it, yielding promising new opportunities for effective, genetic-based treatment. Gene therapy delivers therapeutic substances directly into a patient's cells to treat a disease—or, in this case, to treat a vulnerability to a disorder.

The next step is to confirm these findings and try such a genetic-based approach in a group of war dogs. Again, it would be the right thing to do for the Department of Defense to step up.

Recently I attended a meeting of the American College of Veterinary Behaviorists. One of the talks was given by my colleague Dr. Walter Burghardt, a war dog specialist. A veterinary behaviorist like me, he has served as chief of behavioral medicine at the Daniel E. Holland Military Working Dog Hospital at Lackland Air Force Base in Texas and is responsible for the behavioral care of more than 1,500 military working dogs around the world. He is considered an authority on what he refers to as C-PTSD (canine PTSD), which he defines narrowly. The added *C* in the acronym is designed to calm the kind of uproar that developed over Gina's condition by a name some people thought should be exclusively reserved for *Homo sapiens*.

Dr. Burghardt describes himself as an "extreme behaviorist," meaning he observes but does not interpret behavior. To be diagnosed with C-PTSD, dogs have to have a history of deployment in a combat environment. Second, they have to have been present at a "combat event." Dr. Burghardt's clinical signs of the condition include: hypervigilance to environmental events, escape/avoidance reactions when faced with certain challenges, generally skittish behavior (tail down, acting scared all the time), changes in social rapport (increased dependence on the military handler), and disruption of task performance.

The onset of these behaviors would have to occur during or after a combat event and persist long afterward for the diagnosis to be confirmed. He found that the prevalence of C-PTSD was 5–10 percent of deployed dogs, making it a serious concern for the military. Curiously, Labrador retrievers were the breed most affected.

Dr. Burghardt's assessment of canine PTSD is essentially similar to mine, although his definition is a little more narrow. He deals with war dogs and has little cause to be concerned with what the military calls civvy street. I consider the Army's approach to *treating* C-PTSD to be a tad conservative and only somewhat effective. We could better serve afflicted dogs by funding an array of research into optimal therapies. We could also confirm the genetic cause for susceptibility, thus allowing the military in the future to choose dogs for military service that were almost literally "bomb proof."

Perhaps some of these future research findings would translate into new possibilities for human war fighters, too, opening up better training, deployment selection, and treatment options. And isn't that the bottom line?

CHAPTER FOUR

The Emotional Wounds of Rescued Animals

Compassion for animals is intimately associated with goodness of character and it may be confidently asserted that he who is cruel to animals cannot be a good man.

—ARTHUR SCHOPENHAUER

Comet, Poker's Queen Bee, Mobey, Elsa, Gina—so far we've put more than a few animals "on the couch" for analysis. But I have gotten a little ahead of myself, and I should fill you in about how I came to treat behavioral issues in animals in the first place.

When people find out what I do for a living, when they learn what my entire life's work has been, they usually assume that I've had my own pets. And they're mostly right. I've always lived with easily looked-after pets like cats, rodents, and birds. And I have horses in full board at a nearby barn. But because of my long hours at the veterinary clinic, I couldn't make enough time to properly care for a dog. In recent years, with a more flexible schedule, even though I didn't really set out to become a dog owner, I've been lucky enough to share my home with two wonderful rescue dogs, Rusty and Jasper.

First, there was Rusty. One summer afternoon, as I was about to tee off on the golf course, my wife, Linda, who is also a veterinarian, called me from a local shelter, the Baypath Humane Society of Hopkinton, in Hopkinton, Massachusetts. Because our kids were grown and gone, Linda and I had been in search of a cat or two to rescue.

The old saying that owners are picked by their dogs, rather than the other way around, applied in this instance. Upon arriving at the pound, Linda spotted a dog being walked across the parking lot. Rusty spotted her at the same time. She knew immediately that he would be ours.

Since I was about to tee off, I had to trust Linda's judgment. The same day, she also managed to rescue a cat. You know, in order that Rusty might have company!

What could possibly go wrong? A veterinary behaviorist and his house-call veterinarian wife had rescued a dog from a local shelter. Let's just say it came as a surprise to find out, very early on, that Rusty had some significant behavioral issues. At least he'd wound up at the perfect home. It was like a chocoholic getting adopted by Willy Wonka.

The first sign of Rusty's distress came when I arrived home from work one day and removed my tie. That simple gesture caused Rusty to cower, which he would also do any time I removed my belt. Cowering wasn't the only response. Rusty also leaked. A submissive urinator, he displayed overwhelming and unnecessary groveling behavior whenever I greeted him. He wasn't just glad to see me. He couldn't contain his joy and in his desire to be submissive to my "almighty" towering presence, he wetly demonstrated his response all over my shoes.

The flip side of this anxious conduct was significant separation anxiety. If Linda was away in the barn, Rusty sometimes took that as a cue to defecate in the house. Basically, he was so stressed that he had an accident in his pants, even though he wasn't wearing any. Rusty also liked to empty our garbage can if we left him home alone even for a short time. Yet he reacted with terror whenever we shook out a new trash-can liner.

All in all, Rusty faced a long road back to something like normal. Fortunately, Linda and I had strategies with which to help him.

The first such measure was to make him feel safe. He'd obviously been harshly disciplined at some point in his short life. He might have been beaten, presumably with objects that resembled ties or belts, which could explain Rusty's reaction when I removed my own belt or tie. He could also have been flashing back to earlier mistreatment or punishments when he had accidents in the house or emptied the trash can.

Linda and I put in place a new set of house rules. We would not verbally scold Rusty whatsoever; we would not confront him about his accidents or other anxious behavior, and of course, we would never physically punish him. We wanted to build up Rusty's confidence, and then we'd see just what kind of dog he would be. Up to now, he was fearful. We looked forward to discovering what kinds of personality traits a non-anxious Rusty would exhibit.

Our work took a while to take hold. And it was work. It's difficult to restrain the very human impulse to become snappish in response to misbehavior, but we did. And slowly, gradually, Rusty came around. Linda and I were re-ordering his world, changing it from one of fear and anxiety to one of trust and safety. The incidents and accidents decreased until they disappeared entirely. In place of Rusty the Trembler we had now teased out the dog he really was, Rusty the Keeper of the House. He loved his new place in the hierarchy and he knew his important role.

As a researcher into animal behavior, I think about all the curses, shouts, and screams that are flung at dogs, the rolled-up newspapers, the paddles and straps. It's all so unnecessary. Much of the time the dog simply serves as an outlet for a human's frustrations. Rusty demonstrates that a careful, reasoned approach to correcting unwanted behavior in dogs is effective. Emotional, irrational approaches often are not.

Some people have to go through years of study to understand such basic truths; others come to it naturally and intuitively. I was blessed to be introduced as a child to the wild, woolly, and wonderful world of animals by my mother, who had enormous reservoirs of empathy. I grew up in a household where animals held a central place. My mother was my first teacher and set me on my path in life. Because of her influence, I've always been immersed in the animal world, watching how they function, examining their behavior, and marveling at how like us they can be.

Gwendoline "Gwen" Dodman was a wonderfully warm individual. People loved her, and other creatures did, too. She was always patient, never angry. I have visions of her during my childhood, smiling, lightly built, almost thin, very pretty. Her blond hair was so long back then that she could sit on it when playing the piano.

My mother's twin delights were music and animals. By the time she was twelve, Gwen was teaching other kids to play the piano. Shortly thereafter she was asked to perform on BBC Radio, back in the days when radio was king because there was no TV. Animals were her other passion, especially birds but really all creatures great and small.

My own passion started out very, very small. Even though I encountered dogs, horses, and cows growing up, the earliest animal I remember, and the most prominent, was a bird the size of an infant's fist. Pippit was a fledgling song thrush who had the misfortune to have fallen out of the nest near my grandmother's front porch. Though the chances of a bird thriving after being put back into a nest are low, my mother felt we needed at least to try. She found the nest out of which the bird had fallen in a nearby yew tree and lifted the baby back into its home.

I was eight years old at the time. The nest was within view of my grandmother's window, from where we monitored the situation for a full day. I sat anxiously for long hours, staring, hoping, waiting for the mother thrush to return. The nest looked strangely quiet, though, so we returned the next day to check on the little bird.

When I scaled the yew tree and got a glimpse, the baby bird was still there, but she was clearly alone and neglected. After returning to the house and spying on her for a while longer, we could tell that no mother bird was coming to the rescue. I begged my mother to intercede. She reluctantly agreed that the only possible course of action was to bring the barely feathered, broad-beaked nestling back to our home and attempt to preserve her young life. My mother gently scooped up the bird and put it into a shoe box that she had feathered with a mixture of dried grass clippings and moss. The creature resembled a pint-sized space alien, goggle-eyed and gawky. We stared at each other, bird and boy, my eyes as wide and bulging as the near-naked being in the box.

The baby thrush was no doubt overwhelmed by the smell of its new shoe-box home, formerly occupied by a pair of leather brogans. Mother tightly applied the cardboard lid, which she'd already perforated to allow proper ventilation, and together we set off from my grandmother's house for home.

The days and weeks that followed left such an impression on me that I can trace my belief in the interdependence of humans and animals to that time. At first, my mom fed the bird every two hours. She had cheerfully concocted a home-style chow made of worms and bugs, a process that alternately fascinated and disgusted me.

Once we were sure the bird would survive, and not until then, we gave her a name.

"Pippit," I would whisper, monitoring the creature as it wagged its out-of-proportion head, cheeped pathetically, or slept like the dead. For some reason, without any evidence one way or another, I decided the little slip of a thing was female.

At feeding time Pippit would hunker down, point her head skyward, open her beak widely, and flap her tiny wings as she hungrily gulped down the spongy-textured chow. The little bird grew and grew. Eventually, we were privileged to watch as she attempted to fly, flapping her way across the carpeted living room floor. After a few days she was able to fly short distances. Then there was no stopping her. She flew to perch on top of everything, including my mother's mahogany Bechstein piano, where she'd merrily play with a rubber band or some buttons.

Pippit was loved by all of us, my sisters and me, and my parents. I believe she thought of my mom as her mom. But it came time to return her to her natural habitat. It was only right to let her go. With mixed emotions, we opened the French doors to the living room and, after some slight hesitation, Pippit flew off.

For weeks, I was inconsolable.

Then, a year after I had said good-bye to the little orphan bird, something magical happened. Spring had come, and even though it was rainy old England, we were able to throw open the French doors again to get a much-longed-for breeze into the house.

On that breeze and straight into our living room came a bird that looked just like Pippit.

We weren't sure at first. When the creature headed straight for the piano that was Pippit's usual resting place, we knew that this was, indeed, our dear bird come back to visit. Perching there, as she had done a year earlier, she happily turned her head from side to side, alert, and seemingly once again at home. Out in the garden on the fence, another thrush sang and twitted. We assumed it was Pippit's mate, waiting for his girlfriend but not tame enough to come inside.

And that's how Pippit told us that she had survived and made a life for herself. After a few minutes, Pippit joined her mate out in the garden and left for good. But the effect on me was profound and echoes to this day. A creature had flown across the species barrier and touched my mother and me. Later on in life, I became all about challenging that barrier, reaching across it, questioning whether it is really a barrier at all.

I grew to know many other animals as a child. My mother delighted in rescuing all manner of beasts. She was the neighborhood St. Francis of Assisi. It seemed like every few weeks someone would turn up on our front doorstep with some little furry or feathery creature in need of attention. Often I'd find some small animal in a box in the kitchen or laundry room. Although birds had a special place in her heart, Gwen cared for whatever species turned up, going to great lengths to make them comfortable and healthy. Worm chow was a constant of my young life—for the birds, that is.

Her compassion for animals rubbed off on me. I grew up believing her empathy and kindness was perfectly normal, the rule rather than the exception. To her, animals were sentient beings with feelings and emotions much like our own. They felt pain, they could be frightened, and they could become defeated and depressed or feel happiness, joy, fulfillment, and contentment. I still feel the same way about animals. When I look now at Rusty running after a tennis ball in my local park, for example, I see a once-frightened dog with behavioral issues transformed into a perfectly content, happy, and playful animal.

But Rusty and his issues were to come into my life much later. First, I would follow my calling and head off to university to study to be a veterinary student. On vacations I would return home and see my mother working in the kitchen on a summer day, door open, with wild birds hopping around on the counters or perched on her head. Even though my days were taken up studying animals, it was always a shock to be so suddenly reminded of my roots.

My mother's approach was simple. Win the trust of animals first, then live with them and find her place within their group. She had little patience for the theoretical data of psychologists or ethologists, those who simply observed from the sidelines. As she saw it, such scientists considered animals to be social robots with hardwired behaviors, so-called "fixed action patterns," which could be triggered by specific "sign stimuli." Viewed this way, animal behavior was about as warm and fuzzy as the workings of a grandfather clock. The experiences of Jane Goodall, Biruté Galdikas, and Dian Fossey, living and working within the families of chimpanzees, orangutans, and gorillas, were much closer to my mother's experiences with her birds and other animals.

But the beliefs Gwen Dodman passed down to me are not always shared by others in the veterinary field.

Following the path I had first started on during my childhood, I chose a course of study in school that would have me caring for animals. I qualified as a veterinarian. From my mother and some good experiences as a veterinarian student, I came to believe without a doubt that empathy and affection for animals goes a long way toward making a good veterinarian. Yet I found that views to the contrary abounded in science and in the vet business.

A professor of human anesthesiology once said in a lecture that he did not believe animals feel pain. "Pain," he pronounced, "is a subjective moiety whose perception requires interpretation in highly sophisticated brain regions." I wanted to raise my hand and suggest that these regions in animals are much like those in humans, but the man went on, airily suggesting some form of species noblesse oblige: "However, I strongly believe that animals should always be given the benefit of the doubt regarding pain management and be treated as sentient beings."

Even though he disavowed the whole notion of pain in animals, that same professor clearly acted on that "benefit of the doubt." An owner of two gorgeous Golden retrievers, he freely admitted that if either dog ever needed surgery he would insist on full anesthesia and pain medications. Theory and practice don't always jibe.

Setting aside differences in philosophy, it's a sad fact that not everyone is kind to animals. Some enjoy inflicting pain and suffering. People who behave cruelly to animals usually treat their fellow human beings badly. This has been proven time and time again, and it's the sorry corollary to my belief that all living creatures, human and nonhuman, deserve the same treatment and the same medical care.

Once I qualified as a veterinarian, I specialized in veterinary anesthesia. When I was still perfecting my skills, I was regularly invited into hospital operating rooms to see the practical side of *human* anesthesia and surgery.

One day I stood in a prep room outside the OR, masked up, waiting to watch the induction of anesthesia in an already sedated elderly woman.

"Would you like to do this one?" one of the anesthesiologists asked.

"Okay," I whispered, "but won't she object to having a veterinarian work on her?"

"If you keep your mask up, she won't know," my teacher replied.

So there I was, administering a dose of barbiturate intravenously as the lady breathed oxygen via a face mask. I chased the barbiturate with a small dose of the muscle-paralyzing drug, succinylcholine, which back then was standard procedure, though nowadays other relaxant drugs are more commonly used. The woman lost consciousness and, as happens with succinylcholine, her muscles began to contract and twitch. Orderlies held her arms down to contain her movement until she became completely paralyzed. I then peered down her throat using a laryngoscope, slid a tube into her trachea, inflated the cuff, and hooked the tube up to a breathing bag. I rhythmically compressed the bag as the gurney was wheeled into the operating room.

Once there, she was hooked up to a ventilator by the tube in her trachea and the anesthetic vaporizer was pushed. As one final fail-safe, the woman was given a longer-acting muscle relaxant to suppress any reflex movements she might have during the operation.

"That wasn't so bad, was it?" the head anesthesiologist asked.

I had to admit that it was easy. In the course of my studies, I had anesthetized dogs, cats, horses, cattle, goats, sheep, birds, monkeys, some big cats, a tapir or two, pigs, and, once, a potoroo, a rabbit-sized Australian marsupial. Now I had added a human to that list. This just wouldn't have been possible if all of us animals weren't so similar physically.

In 1981, after some academic twists and turns, I found myself in the United States, at Tufts. During my first few days in Boston, I took a trip up the sixty-story John Hancock Tower on a crystal clear day and surveyed the sprawling metropolis before me. Outside the heavily populated center, the city landscape melded into other, smaller towns. The horizon appeared far away. From my skyscraper aerie I could see the coastline fading into the distance.

This was my new territory. There wasn't a veterinary anesthesiologist as far as the eye could see. It was an exhilarating thought. I was a pioneer in New England and at the same time a pilgrim of sorts.

———————

What I liked most about my new home at Tufts was the philosophy of our then president, Jean Meyer (pronounced in the French way, JHON My-EER). Presiding over the medical, dental, and veterinary schools, Meyer had developed a concept he called One Medicine. It was the first time I encountered the term that would become the fundamental underpinning of my career.

The gist of Meyer's thought is that human medicine, dentistry, and veterinary medicine are all variations on a common theme. He encouraged cross-campus liaisons and cooperative research, which brought together specialists from the various schools with different takes on the same underlying theme. I needed no special coaxing to fully sign on for that approach to treating animals.

It's not rocket science, as they say. The fundamental facts are easily grasped and not really challengeable. Dogs, cats, and other mammals have brains that are very similar to our own brain in both structure and function. But it's not just brain anatomy that is similar across the species. Animals in general, and mammals in particular, have many of the same inner workings as humans do. We share physical similarities and we respond to incoming sensory information in much the same ways. Under the hood, so to speak, in terms of the nervous system or other organ systems, there is not much difference in how things work.

For this reason, domestic dogs and other animals have served as experimental models in psychological and neuroscience experiments for centuries. Ivan Pavlov examined conditioning by ringing a bell at feeding time, which caused his canine subjects to salivate even if they got no food. B. F. Skinner established the existence of operant conditioning in hungry rats, when behavior rewarded with food was repeated while behaviors that were not rewarded died out.

Today, while many people accept that nonhuman animals have the capacity to think and experience emotions, some still have trouble accepting that animals have a sense of self, or "theory of mind." Theory of mind implies that individual animals can comprehend that other beings have information that they don't possess, and that others may have different and competing desires and aims. This sounds completely plausible to me.

A now-famous experiment by Dr. Brian Hare illustrated that dogs clearly understood that the person pointing the finger knew something they didn't, when the dogs followed a pointing finger to find a food treat hidden under a bucket. The canines employed the human's know-how to find the treasure. Dogs also show that they possess desires and intentions when they hide a bone or food treat from competitors.

We need to accept that species other than *Homo sapiens* have theory of mind in order to believe that animals experience secondary emotions like affection, contentment, and suffering, as well as tertiary emotions like eagerness, frustration, jealousy, and guilt.

Let's take as one example what Shakespeare calls "the green-eyed monster" of jealousy, which is supposedly specific to humans. Dogs who push in between embracing couples or between their owner and another dog demonstrate jealousy. In a recent study by Christine Harris, a professor of psychology at UC San Diego, dogs exhibited significant unhappiness when their owner paid attention to a stuffed toy that barked and wagged its tail.

In my work at the clinic, I see dogs exhibit jealousy on a regular basis. I am very comfortable calling it like it is. Brody, a six-year-old neutered male Yorkshire terrier mix, is a pocket-sized Napoleon of a dog who would take up residence on the couch next to his owner. He would not let anyone approach without barking and lunging, which usually had the desired effect of making the person back off.

If his owners sat on the couch together, Brody would make every attempt to insert himself between them. He made his displeasure at their closeness very clear. In the evening, the dog favored his male owner. The wife was often forced to move. God help the couple if they were to touch or embrace. That was definitely not allowed. During the day, Brody transferred his allegiance to his female owner, threatening anyone, family or guests, who approached her. Brody wanted her exclusively for himself.

Another Yorkie I knew would hop into bed with his lady owner each night. He'd stand on alert on the husband's side of the bed, waiting for him to come upstairs after turning the downstairs lights out. If the husband dared to climb into his own bed, the Yorkie attacked. More than once the husband found the jealous reprobate firmly attached by the teeth to one of his fingers. For this husband, discretion was the better part of valor. He often gave up the field and crawled off to sleep in the spare room.

Dogs may also be jealous of one of the other canines in the home. They will do everything they can to break up any attention shown to it. Jealousy describes the situation perfectly. Increasing evidence supports the contention that animals have more than merely a vestige of secondary and tertiary emotions.

I've learned a lot from treating animals, but I've learned a lot about resilience from my own pets. How our Rusty overcame his fears represents just one example. My other dog, Jasper, survived neglect before he came to me. As I reach down from my writing desk to give him a pat, I find once again the lesion on his upper back that was caused by a long confinement in a crate that was too small for him. The lesion is now superficially covered with beautiful, healthy fur, like a comb-over. But the lesion is still there, and it always will be.

Some animals are born with genetic tendencies to develop behavioral problems. Others have behaviors forced upon them by bad owners. Jasper, who otherwise could have been a normal dog, arrived at my home with issues because his earlier owner had treated him poorly.

Jasper's original owner was a health-care professional who worked with autistic children. Her job caused her to be away from her apartment for long periods. As a result, Jasper, a mixed-breed, long-haired beauty, found himself in a crate for much of every day, sometimes for as long as twenty-four hours at a stretch.

The crate was so small that Jasper could not stand up to his full height. If he ever barked when out of the crate, he was put straight back in. When he was taken outside, he was quickly walked and then re-crated. The result was that, as soon as he left the building he let loose. He peed literally as he stepped outside, because he never knew how long he'd have.

His miserable life might have gone on like this forever but for the intervention of my daughter Keisha. She found herself in the apartment where Jasper was crated, and she couldn't believe what she was seeing. Keisha called me to fill me in on the condition of the dog and wondered if I or her mother could help him.

We were not looking to add to our collection of pets, a late-in-life menagerie that had grown. We were still training Rusty to turn him into a calm, non-frightened dog. Rusty already had a pal, Griswald the cat, the one who came into the household at the same time. Griswald was quite pushy, demanding to be petted by rolling on his back and looking cute, whereupon he would then bite the hand that petted him. He was deaf but was also very vocal, able to sing like a feline Caruso.

In Jasper's case, there was no mechanism for us to intervene. The dog belonged to its owner. Nonetheless, as I listened to Keisha's tale of woe, I felt my blood beginning to boil. Pets are helpless. They deserve humane treatment. But in this case, I was the one who was helpless. I wasn't in a position to assist this poorly treated dog.

Keisha continued to monitor Jasper's situation as much as she could. She did her best to get him out of the crate whenever she found herself at his owner's apartment. On one visit, she judged that Jasper's situation had become dire and called me in a panic.

Jasper had eaten something nonedible. The poor creature had made a meal out of a whole box of tampons. It wasn't surprising the dog had ingested what he shouldn't have, since he had become emaciated from being confined. He weighed forty pounds instead of the nearly eighty he should have. His appetite for both the edible and inedible was voracious.

He'd vomited up a few tampons, but it was clear that he had consumed many more and would require surgery to remove the others that were still in his system. The price tag for such an operation was a cool three thousand dollars.

Jasper's owner didn't have that kind of money. Even if she had, it was clear that she didn't know how to care for a dog, healthy or not. Linda and I discussed what to do. Keisha was distraught. The dog faced either an expensive surgery with a return to poor conditions or euthanasia. Linda then suggested that we offer to do the surgery gratis. But we'd also keep the dog. The harder task was to convince Jasper's current owner to go along with the plan. It would prove harder than we imagined.

The woman remained in denial about the bad shape Jasper was in, but after my wife and Keisha and I had a difficult and emotional conversation with her, we came to a mutual agreement that we would perform the surgery and keep the dog. Finally tearful and repentant, Jasper's owner dropped him off at our practice and said her final farewell.

Linda did the surgery, while I was in charge of anesthesia and giving the dog intravenous fluids. The tampons had lodged at the outflow end of the dog's stomach. Jasper is deep-chested, so they were not easy to get out, although eventually Linda managed it. She removed one via a stomach incision, and another she massaged so that it traveled farther down into the large intestine, from where Jasper could easily pass it without fear of further obstruction.

After the procedure, Jasper gradually woke from the effects of anesthesia. The pooch had no idea what had just happened. He didn't know we were to be his new, more attentive owners. He just understood that he felt a whole lot better.

It's an amazing process, to see an animal return fully to life. Over the weeks that followed, our new housemate regained strength. He learned to trust us. He soon got back to a normal weight of seventy-five pounds and was no longer desperate to eat everything in sight. Food wasn't something to be snatched from the conveyor belt of life and he learned not to fear losing every fleeting opportunity to eat. He became much less fanatical about his chow and would occasionally leave some food in the bowl. Nor did he practically take our fingers off when we offered him a treat.

After a few short weeks, Jasper also realized that a walk outside was not going to be over in thirty seconds. He no longer peed on the back patio two steps into an excursion. He enjoyed morning romps in the woods behind our home and two long walks, both off leash, a previously unheard of privilege.

Linda and I monitored him as his muscles came back and regained tone. He got a spring in his step, a smile on his face, and learned to love swimming in nearby lakes. He was no longer crated, ever. He and Rusty became brothers in arms, a virtual Huck Finn and Tom Sawyer duo, though I would be hard put to assign which role to which dog.

Jasper now sleeps wherever he wants, which is usually in bed next to me at night or on the couch in the early evening. I often awake to find his head on the pillow next to me, his eyes fluttering half-shut, and with his long nose an inch or so from my face. He's tired and happy by the end of each day after his many walks, but even with these outlets is ferocious with chew toys, able to take apart even the most durable ones in minutes. For this quality, he has earned himself the nickname of Jaws of Steel.

One other thing became evident as Jasper blossomed. As Big Mama Thornton would sing, he "ain't nothin' but a hound dog." He spends half his life with his nose glued to the ground. Once he's caught wind of something in the woods, he's fast on the trail. At these times, he suffers from an abrupt, temporary, and, I suspect, voluntary hearing loss. No matter how much I call at these times, he simply ignores me and goes about his all-important mission.

This is what we want for all creatures, human or otherwise. We wish them to fully inhabit their true selves. In his previous crated life, Jasper's potential was pinched back to almost nothing. Now his "walk-about" excursions have become so far reaching that I have had to fit him with a GPS tracking device. Even years after we took him in, his boldness increases by the day but I don't want to rein him in.

CHAPTER FIVE

The Dog Who Couldn't Stop Licking

Compulsive Disorders

May all that have life be delivered from suffering.

—BUDDHA

Even if you haven't personally had experience with pets who have licked, chewed, or rubbed themselves raw in places, you might have seen it in other people's pets. The condition can appear ghastly. The natural impulse is to look away.

Most dogs that lick themselves raw have allergies, but there is also a condition called lick granuloma, a very specific type of compulsive behavior that is directed only at the lower extremities of the limbs, above the foot but below the elbow or knee. These areas are termed acral, a medical term for "affecting the extremities."

For a long time, back in the Dark Ages before researchers began to unlock the secrets of animal behavior, vets and owners alike treated all irritated, furless patches on dogs with salves and ointments. These treatments can help sooth allergies, but with compulsive acral licking, they miss the point. Lick granuloma is not disease of the skin. It is a problem of the mind.

After Dr. Lou Shuster and I investigated compulsive behavior in equine subjects like Poker's Queen Bee and Mobey, we wanted to know whether our findings in horses could apply to other animals.

So we turned our attention to dogs. Although the veterinary profession's roots are in cattle doctoring, in today's world canines have pride of place in examination rooms and animal hospitals. There are 78 million dogs living in American households. Lou and I asked ourselves, What behavioral condition do dogs have that are like cribbing: repetitive, apparently mindless, and seemingly pointless? In other words, what ways did dogs display compulsive behaviors?

Lick granuloma immediately came to mind. In response to constant licking, the skin erodes. In the normal process of healing, tiny blood vessels form and new connective tissue gradually fills in the wound. This healing process is one of the wonders of life—in humans, dogs, and other animals. But in lick granuloma, this process goes awry producing exuberant mounds of pink granular connective tissue bulging from the sore. This happens because the affected dog does not allow the healing process to run its natural course, and constantly irritates the wound by the licking. Sometimes, a deep ulcer forms and infection can track deep into the underlying tissues, including bone.

Lick granuloma is also referred to as acral lick dermatitis. The location of the licking and resulting abrasion is fairly specific and usually on the wrist (foreleg) or hock (hind leg). These are the locations where the dog's head comes to rest if he tends to lie down flat on his chest, paws in front of him, or curled around on his side with his chin resting on his hock.

To test our theory that lick granuloma was natural behavior gone wrong, Lou Shuster and I repeated our experiments using morphine antagonists on dogs who had various forms of self-directed licking, scratching, or chewing. The dogs responded as the horses did—they reduced the harmful behavior. Clearly the drug's action—and the underlying biochemical mechanism—was not just a "horse thing." We then used the same drugs on cats who chewed blankets, stripped out their hair, or had other relentless neurotic habits. The cats, too, responded well.

At the time we were involved in our research, the behaviors we were treating were considered stereotypies—that is, rhythmic, repetitive, fixed, purposeful movements that occur in an animal that had otherwise developed normally.

But to call acral licking a stereotypy describes the symptom—yes, *of course* it's a repetitive act!—and does nothing to get at root causes. Back then, the behaviors were not considered compulsive disorders. No one knew what triggered them and few vets knew how to treat them.

Acral licking is painful to watch, and many owners of pets with the affliction react with the same kind of frustration triggered by a cribbing horse. They take it personally and go to vets in despair. "How can I get my dog to leave off this constant, maddening, stop-it-already-will-you! He's just doing it to annoy me!"

Many vets tried, but standard approaches of the day didn't work. Dogs and cats were fitted with "Elizabethan collars" to prevent them from licking or chewing. These triggered a lot of blundering into furniture and perhaps some embarrassment in the pets, but upon removal of the collars the behavior recurred. Vets sprayed the affected areas with substances like bitter lemon that tasted or smelled bad, to try to repel the dog from licking. They injected an analgesic and corticosteroid combination to numb and soothe the injured skin areas, thinking that the dog would then not be so conscious of pain or irritation in the affected area, and thus not be prompted to lick.

The condition so frustrated the profession that it was turning vets into modern-day carnival snake-oil salesmen. Some vets applied venom milked from the king cobra to numb the area and relieve pain. This was not obtained humanely from the snakes, and it did not succeed in curing the discomfort of the pets treated with it. Other vets tried "Substance P," an extract of the pitcher plant with long-acting local anesthetic properties, which also failed to help.

These approaches assumed that, if only the lesion could be numbed, then the dog would stop its licking. And they were all unsuccessful because compulsive disorders have psychological, not physical, roots. I once encountered a dog who had no sensation in his foreleg as a result of being hit by a car. Despite this lack of feeling, the dog continued to lick a granuloma lesion on the same foreleg, even though the nerve supply to that leg had been demolished. Local pain and discomfort do not cause compulsive licking. Bitter-lemon sprays, cone collars, snake venom aimed to treat the symptom, when we needed to find a way to treat the cause.

In the midst of our studies on stereotypies, Dr. Judith Rapoport, a child psychiatrist at the National Institutes of Health, published a book, *The Boy Who Couldn't Stop Washing,* about childhood obsessive-compulsive disorder. When she returned to her office after her national book tour, she found a slew of messages saying, "My dog does that compulsive self-washing, too." Some psychiatrists might have dismissed these messages out of hand, but Dr. Rapoport took them seriously. What if these dog owners were right?

An expert at designing mental health studies, Dr. Rapoport planned an experiment in which dogs with acral lick would be enrolled in clinical drug trials for three known anti-obsessional medications: Prozac, Zoloft, and clomipramine. She found that the dogs responded to human anti-obsessional medication just as human subjects did, and so Dr. Rapoport concluded that dogs could indeed suffer from a canine form of OCD. In addition, and just as importantly, their compulsion responded to the same medications that help human sufferers.

Following Dr. Rapoport's initial publication on canine OCD in the *Archives of General Psychiatry* in 1992, stereotypies were reclassified as compulsive disorders in veterinary medicine. Her publication was a game changer for all of us vets interested in repetitive behavior disorders. Vets now had a new understanding of what was going on and could develop more logical and successful ways of treating these conditions. These behaviors weren't simply mindless and pointless after all, but were an animal's way of dealing with underlying anxiety and stress.

Dr. Rapoport's insights led me to take what I call an ethological approach to understanding compulsive behaviors in all animals. Ethology refers to the behaviors of animals *while in the wild*. Ethology might be a somewhat confusing term, lost, as it can be, among the welter of "ethnology," "etiology," and the myriad other "-ology" words. But it is a simple concept to grasp. When looking at pet behavior, we recognize the natural conditions that influence how the animals act, and we consider the evolutionary processes that might have led to their adaptive traits.

With Poker's Queen Bee and Mobey, for instance, their cribbing was a response to being denied their natural activity of free-range grazing. Confinement prevented a normal activity (grazing), which led to an abnormal activity (cribbing). Ethology provides a link between a creature's natural and unnatural behavior.

Dogs are by nature a predatory species. Hunting is how their ancient untamed canine ancestors stayed alive. Hunting has two basic components, the chase and the meal. The two phases of canine predatory behavior are the "appetitive" phase, in which dogs search for, track, and home in on prey—and includes delivering the final killing coup de grace. What follows is the "consummatory" phase. After that, the behavior resets, and the phases repeat.

A common version of canine compulsive disorder involves visually focusing on and chasing elusive prey, whether lights, shadows, objects, insects real or otherwise, or even the dog's own tail. This is the appetitive phase gone wrong. One owner of an English sheepdog described her dog's behavior as "chasing imaginary rabbits around the kitchen until they disappeared down imaginary rabbit holes." A German shepherd I knew of spent a good deal of her day snapping at imaginary flies. The behavior had started with real flies but continued in their absence.

I once encountered a junkyard dog who spent his days looking for something that wasn't there. Amid his vast territory of trashed-out automobiles, the junkyard dog would search and search without ever coming up with anything. It was nonstop and pitiful. This dog was displaying appetitive behavior gone awry, a naturally occurring activity with the brakes off. His behavior recalled the old Hughes Mearns poem:

> Yesterday, upon the stair,
> I met a man who wasn't there.
> He wasn't there again today
> I wish, I wish he'd go away . . .

In these out-of-control appetitive behaviors, what once was natural becomes unnatural.

Consummatory phase compulsions in dogs include flank and blanket sucking. Dobermans, in particular, are at risk for this condition. Dobermans in the grips of this condition can suck their flanks raw. The ingestion of inedible objects can be another expression of consummatory behavior gone awry. Rock chewing, perhaps more annoying to owners than hurtful to dogs, is one example. Normally the affected pooches don't swallow the rocks, but the habit can wear down teeth. In the rare instances where the objects are ingested, they can cause intestinal blockage, a potentially dangerous situation.

Dogs are also a grooming species. Grooming is necessary for a dog in the wild, because it helps remove traces of the last meal. Even a slight scent of decomposing food could be a giveaway when hunting upwind of some flighty prey. In addition, grooming helps keep a dog's fur in tip-top condition. A dog that did not self-groom would be more likely to succumb to infections. Evolution selected out careful groomers for survival. But as we've seen, overgrooming leads to acral licking, the prototypical canine compulsion.

Other survival-linked behaviors include digging, an activity that is quite useful in the wild and harmless in pets—until it goes off the rails and becomes compulsive. Swimming is obviously natural, especially in water breeds, such as Chesapeake Bay retrievers, Newfoundlands, and Portuguese fishing dogs—President Obama's favorite breed. But obsessive swimming, which can occur in these breeds, can lead to exhaustion and disruption of the so-called human-animal bond.

Lief, a young German shepherd, was driven to distraction chasing his own tail. I'm not talking about a few desultory backward lunges from time to time. I'm talking about frenetic spinning in tight circles for hours on end. Lief was totally fixated on his tail. And he was becoming gaunt from his continuous, frantic spinning and barking.

Anxiety appeared to be at the root of his problem. Hypervigilant by day and by night, Lief would whine and chase his tail at the slightest provocation. A common trigger was the sound of passing cars at night as well as their headlights making fast-moving shadows on the ceiling. Some anxious German shepherds will bite and bloody their own tails, to the point that the tails have to be amputated. Lief's problem was close to being that severe. He had become so anxious that he barely ever made eye contact with his devoted owner. His behavior was not in any way natural or functional. It was pointless, repetitive, and verging on self-mutilation.

Attempting to retrain Lief so that he did not chase his tail would not have made any difference. He had a psychological problem, not merely a behavioral one. His tail chasing was likely the misdirection of pent-up, predatory urges. I diagnosed him with canine compulsive disorder.

But tail chasing in German shepherds is sometimes complicated by a localized kind of canine epilepsy. In people, OCD has long been linked with epilepsy.

We were able to control Lief's problem using an anti-obsessional medication. Once he calmed down somewhat, we adjusted his lifestyle and ensured that his owner changed his own behavior in order to create a more entertaining environment for Lief, which would make him less anxious. He gave the dog more exercise, more attention, and more fun activities. We also added an anticonvulsant to the medication regimen, to address the suspected underlying seizure activity. Lief improved—"a night-and-day difference," according to his owner. He lived to a good old age and died with a healthy, well-furred tail still attached to his body.

Whether tail chasing in German shepherds is a purely compulsive behavior or is driven by underlying seizure-like activity in the brain remains unclear. In fact, it's a bit of a chicken-and-egg problem: Does the compulsion cause seizure-like activity, or do partial seizures, localized to a particular region of the brain, sometimes manifest as compulsive behavior? But at least we now have a better handle on managing compulsive behaviors in animals.

Cats also display compulsive traits. Two common feeding-related compulsions in cats are wool sucking and pica, which involve sucking on or eating nonnutritive substances. Wool sucking is similar to the blanket-sucking problem of Dobermans and may be akin to thumb sucking in children. These all represent displaced suckling behaviors. Affected cats will typically target woolen materials, plastic shower curtains, shoelaces, and the like. The behavior can wreak havoc on an owner's belongings, and it is dangerous for the cat, who may wind up with an intestinal obstruction.

Besides cribbing, which has its roots in a pressing need to constantly chomp on something, horses can develop other compulsions related to their free-ranging impulses. They may pace, stall-walk, or fence-walk. Weaving represents an abbreviated form of walking to and fro, in which the horse swings his head from side to side as he walks in place.

Zoo animals exhibit compulsions that are typical to their species. Elephants who roam large distances in the wild will weave when tethered by leg chains. They might walk in place while swinging their heads from side to side. Giraffes do the same. Captive bears pace aimlessly in their cages. Big cats can wear out pathways around the periphery of their enclosure as they stride away their captive lives.

Over the years, I have consulted with zoos about numerous strange animal behaviors, including compulsive pacing behavior in polar bears. In nature, polar bears roam for miles in search of food on the ice shelves. When constrained in the zoo environment, mounting frustration at the monotony of their small environments can cause a polar bear to constantly pace to and fro along the same route.

At the Calgary Zoo, I helped placate a compulsively pacing polar bear named Snowball. The treatment was relatively simple. We gave Snowball Prozac. It worked so well that the zoo staff aired the results proudly in a letter to the AVMA journal.

Gus, a popular polar bear in New York City's Central Park Zoo, like Snowball, exhibited compulsive behavior by a repetitive daily swimming schedule. Visitors to the zoo became upset witnessing Gus's unerring to and fro. News reports made the "bipolar polar bear" into something of a cause célèbre. Zookeepers treated Gus with Prozac, too, although I was not the one to prescribe it. Additionally, his keepers worked to provide him with distractions. They scattered beach balls and other toys around his tightly limited environment—which was, as the press delighted in pointing out, the size of an average New York studio apartment. Gus's repetitive behavior diminished. He died in 2013 at the ripe old age of twenty-seven.

Nearer to my home, at the Franklin Park Zoo in Boston, I came across an unusual OCD-like condition in a gorilla. The vets there described their charge as bulimic. He would eat a hearty meal and then promptly throw it back up. All medical causes of vomiting had been ruled out. But the gorilla was described as anxious, bordering on neurotic, a description used not just by tabloid journalists but by professional primatologists.

Bulimia is thought to be a condition in the obsessive-compulsive spectrum of diseases. The gorilla displayed an obsession that involved the feeling of being overly full after eating, followed by a subsequent compulsion to purge. I recommended treatment with an anti-obsessional drug, clomipramine, and an antianxiety drug, Valium. Some weeks later the gorilla had ceased vomiting and seemed more composed. His condition may have been simple anxiety/stress-related vomiting. It was tempting to place it somewhere along the OC spectrum, perhaps in a new category, the gorilla version of the disorder. GOCD?

In the wild, birds are natural preeners. When stressed by confinement, birds overgroom and pluck out their feathers. Alex, the celebrated African gray parrot trained by the psychologist Irene Pepperberg, was a confirmed feather plucker. Recognized as an Einstein among birds, Alex led something of a high-pressure existence. Between intense training sessions by Irene or her students, Alex had to endure long hours in his cage. Effectively in school every day, he was caged when not being taught to recognize objects, understand concepts, and communicate. Irene also traveled a lot, doing presentations about her prize pupil's accomplishments.

Alex's life was unnatural for a parrot. It's not hard to see why Alex displaced his frustrations as feather plucking, which then became ingrained as a true compulsion. Even Einstein might have become a bit of a nutty professor in response to such limiting conditions. Alex died prematurely from arteriosclerosis, or hardening of the arteries, a condition which in humans is often exacerbated by stress.

Feather-plucking parrots perform a set of ritualized actions that would be fascinating if they weren't so painful to watch. The birds search for new-growth feathers, pull them out by roots with their beaks, transfer them to their claws, and inspect them carefully. They then shred the feather shaft, creating a miniature fan appearance, before discarding it and starting the whole sequence over again.

This behavior is similar to the human disorder, trichotillomania, a compulsive need to pull out one's own hair. Trichotillomaniacs sometimes chew on, or even eat their own hair, a condition called trichophagia. People with trichotillomania often search for new-growth hairs, pluck them out, and inspect them. They might chew the hair bulb. They then discard the hair. The behavior repeats ad infinitum, or in some cases, ad nauseam. The two conditions, occurring in parrots and people, are uncannily similar.

Cats are also "obligate groomers," an easy term to remember if you consider their inveterate need to self-groom. Such natural behavior can develop into an overgrooming compulsion called psychogenic alopecia, which is basically a feline form of trichotillomania. Highly strung, anxious, genetically susceptible cats are particularly prone to developing the disorder when the pressures of feline life become too much to tolerate.

One such cat I encountered began overgrooming while her surgeon-owner was away for five weeks at an international conference. When the owner finally returned home, she found her living room covered with what looked like snow, but was her cat's white fur, pulled out in clumps. Oddly, the cat had begun to bite its nails during her absence, too. The surgeon, feeling guilty that her overseas trip had caused such distress to her beloved animal, brought the pet in to see me. The cat's normal grooming pattern had been exacerbated by stress. I treated the cat with an anti-obsessional medication designed to stabilize its mood. I'm happy to report that it recovered beautifully. The surgeon was grateful. I'm sure the cat was mightily relieved, too.

Another cat—a cute-looking calico called Kalie—developed psychogenic alopecia when her kindly owner, the late Dr. Bob Fleishman, principal of a local veterinary clinic, brought home another cat that had been dropped off at his practice. But here's the rub: cats do not automatically appreciate new acquaintances, let alone permanent housemates, and the arrival of the new cat—well, put the cat among the pigeons (so to speak). Neither do we appreciate everyone we meet, come to think of it. If I was to take someone off the street and drop them off in your home, your reaction would not be, "Oh good, just what I wanted—a fellow human. A person with whom to share my house, bathroom, and dining room." Poor Kalie hid as soon as she clapped eyes on the newcomer; in fact, from that day forth she became a recluse—and started to overgroom and strip her fur out all along her underside and inside her fore- and hind legs. Bob consulted me and asked what he could do. At that time, I was big on the opioid antagonists, so I prescribed one of these medications to be administered once daily by mouth. The good news: it worked! The bad news: the medication was so bitter to taste that Bob could not catch the cat to medicate it after a month or so. It seems the bad taste was worse to the cat than its nemesis, so Bob eventually quit giving it. Poor Kalie had to live out her days privately and somewhat hairless underneath.

Psychogenic alopecia in cats, acral lick dermatitis in dogs, feather plucking in birds, and trichotillomania in people are basically all forms of compulsive overgrooming—this constant repetition of survival-necessary behaviors is thought to be caused by anxiety. Anxiety leads to a "loosening of the brakes" on natural behaviors so that they run amok.

Animals that exhibit compulsive behavior are generally more anxious than their normal, non-compulsive counterparts. To invoke a popular psychological concept, they seem to be worriers with type A personalities. Back when I first began working with anxious, obsessive-compulsive animals, the DSM of mental disorders did not place OCD among anxiety-linked conditions. But the sorting of mental disorders is an ongoing endeavor. A subsequent version of the DSM—"Version IV (Revised)"—changed the classification, grouping OCD with other anxiety-type disorders. OCD was moved again in Version V, and now has a category all its own. I'd argue that this might not be the best placement, given that many people with OCD often do seem to be worrywarts, so the link to anxiety should have been maintained. In my experience, animals with equivalent disorders certainly display increased anxiety.

Anxiety triggers an over-the-top release of survival-necessary behaviors in what develops into an unvarying cycle. Here's how I believe it works: Worry, or indecisive signaling from the decision-making region in the brain, activates nervous pathways "downstream." The performance of a particular behavior pattern lessens the anxiety. In other words, the compulsion serves to relieve the anxiety. But this relief is only temporary. In people with OCD, it's not long before the anxiety and drive to perform the compulsion return. The cycle continues.

No animal in the wild has ever been found to engage in a compulsive behavior. I doubt that humans living in a natural environment, fending for themselves and beset with real worries—let's say, for example, the Korowai tribe in deepest Papua New Guinea—exhibit OCD, either. Cannibalism, yes. Compulsive hand washing, no.

I conducted research in a psychiatric clinic that treated people with trichotillomania, seeking to ascertain when their conditions first started, if they'd ever had any remission, and if they were aware of any triggers. The first patient I spoke with was a man who was a habitual beard plucker (a form of "trich"). The condition had abated once, he told me, for a few weeks when he was hitchhiking across Canada.

This makes sense. On the trip, the beard plucker was looking out for himself, trying to stay safe, sleeping under the stars, making fires, battling the elements, and wondering whence his next meal would come. There was no time to obsess about plucking out his beard as he lived a genuinely natural hominid lifestyle. The behavior resumed as soon as he got home. It took over in force once more in front of a flickering computer screen at his place of work.

The second patient I quizzed was a woman who was a compulsive eyebrow plucker, another form of trich. She had started this odd behavior right after she got married. Her behavior had abated once when she took a cruise without her husband, a round-trip from Boston to the Caribbean. But it had resumed on the way home, four miles out of Boston harbor. Couples therapy might cure her compulsion, but I didn't feel it was my place to make that recommendation.

Animal compulsions that are triggered by stressful events or thwarted natural inclinations may sometimes be relieved by changes in environment. Most compulsions initially arise around puberty or in early adult life, always a stressful time for animals and people alike. For example, around 80 percent of blanket-sucking Doberman pinschers engage in this compulsion for the first time before they are one year old. Sexual maturity in such dogs occurs at around six to eight months, while they are technically still in late puppyhood. Think of adolescent teenagers, and consider their stresses at this stage of life.

Caged mice can exhibit compulsive behaviors, too, though the behaviors they show—jumping, climbing, sniffing, and rearing—are more often referred to as stereotypies. These behaviors can be induced by stimulants and blocked by treating them with an opioid antagonist, such as naloxone. In one experiment, researchers gave compulsively jumping mice two different forms of the drug. One form of the drug was active—effective in blocking the effects morphine and internal morphine-like substances, the endorphins. The other similar drug, because of its mirror-image structure, was supposedly ineffective as a morphine or endorphin blocker and so was used as a control. But to the researchers' surprise, both drugs stopped the compulsive jumping behavior. What on earth was going on? What should have happened, according to our work with Mobey, is that the form binding to opioid receptors would stop the mice from jumping. The other would not.

When Lou Shuster and I looked into this experiment, we discovered that the supposedly inert form of the drug actually did accomplish something. It blocked a set of brain receptors called NMDA receptors and limited the effect of a neurotransmitter called glutamate, which stimulates nerve cells.

Suddenly Lou and I found ourselves playing in a whole new ball game. We were forced to take a new look at our old work. Given what was happening to the mice, the fact that the cribbing horses stopped cribbing could have been due to blocking so-called NMDA receptors for glutamate and for endorphins.

To put our new theory to the test, we obtained various glutamate blockers, some with and some without the reverse opioid structure. We gave one of our horse subjects a very small dose of ketamine, which acts as an anesthetic when used in sufficient quantities. Lou and I settled on ketamine because it, too, is a powerful NMDA blocker. It's also well-known and deadly as a street drug of choice, under the slang name of Special K. One Medicine, indeed.

The first horse to whom we gave ketamine was a champion cribber, just like Poker's Queen Bee had been. When we monitored her before giving her the drug, she chomped at the wooden stall door almost incessantly. She seemed unaware of her environment or us lab-coated observers. After we injected her with a tiny dose of ketamine, the mare reacted with a momentary startle, as if she had just thrown back a double shot of Scotch. Then she relaxed, stopped cribbing completely, and began to show interest in her surroundings.

The drug's effects lasted for around twenty minutes. I offered her a ripe apple that time and she readily took it. Highly palatable foods such as sweet-feed grain and apples can trigger bouts of cribbing in horses, so this was really putting the treatment to the test. But she didn't crib at all. She was just a happy horse munching on a Granny Smith.

To replicate our success, we tried out another, more readily available glutamate-blocking drug, dextromethorphan, on several other cribbing horses, who once again stopped their compulsive behavior. An active ingredient in many cough suppressants, dextromethorphan is known to be safe. We found similar results in dogs who had been self-licking, self-chewing, and self-scratching. We also found that mice with induced compulsive scratching stopped doing it when given NMDA-blocking drugs.

It was a short leap from these findings to concluding that an NMDA blocker might work as a treatment for people with OCD. Realizing the possible value of this discovery, Tufts University took out a patent for the use of NMDA receptor blockers in treatment of compulsive disorders in animals and people.

The patent application extended the human application to the "compulsive components" of tobacco smoking and drinking. Smokers and problem drinkers in remission tend to fight their habit for their entire lives, long after the physical addiction to the chemical has faded away. That resembles a form of OCD. The *obsession* takes the form of daily thoughts about cigarettes or alcohol, while the *compulsion* is the act of imbibing an alcoholic drink or lighting up a cigarette.

The compulsion can be checked with determination and help from support groups, including retraining that substitutes healthy habits for the unhealthy ones. But the obsessive, constantly recurring thought is difficult to suppress. That is why alcoholics in recovery take things "one day at a time" and former cigarette smokers know that even one cigarette might lead them back to their addiction.

We thought that our approach could help quash the compulsion and make recovery from alcohol addiction or smoking easier. Many of the medical treatments currently available do have NMDA receptor-blocking properties and, although the medications are not always billed as working through this mechanism, we believe it to be instrumental in treating these addictions, at least in part by blocking associated obsessive-compulsive components. Pharmaceutical companies, normally eager to exploit every opportunity, did not recognize that aspect of the drugs and therefore paid no attention to our patent. There is still no approved NMDA blocking drug for treatment of animals or people with compulsive disorders despite an estimated $350 to $500 million market.

Dr. Shuster and I needed to find a way to prove that NMDA blockers work for the treatment of OCD in humans. A Harvard psychiatrist colleague, Dr. Michael Jenike, was interested in our idea, agreed to try the NMDA blocker dextromethorphan in some of his most intractable OCD patients, and noted some modest decreases in anxiety for some patients. We then persuaded him to try a more reliable NMDA blocker, memantine, marketed for treatment of Alzheimer's disease in humans under the trade name Namenda. Namenda reduced the OCD behaviors so well that Dr. Jenike still uses Namenda off-label today with many tough-to-treat OCD patients.

When we tried Namenda in dogs with a variety of compulsive disorders, it worked very well. One particular Cavalier King Charles spaniel had demonstrated almost nonstop chasing behavior, fixating on a variety of objects, one after another. His normal canine appetitive activity had gone out of control and he indulged in his compulsion to the exclusion of a normal life. He had virtually no relationship with his owners. If there was nothing else to chase, the dog would throw himself at armchairs to make the dust fly out so he could then chase the particles as they danced in rays of sunlight.

Prozac slowed down this dog, but only the addition of Namenda made the troubling behavior cease.

Not only did our theory prove to be a workable solution for people with OCD, but its discovery highlighted something important about One Medicine. Most often, drugs are developed within Big Pharma companies, tested in rodents, and then undergo human clinical trials until they can be marketed for human use under a brand name. Usually, veterinarians then pick up on the idea and try the same drugs in their patients, often with equivalent success. The pharmacological arrow, in fact, almost always points that way.

There are some notable exceptions to this "humans first, animals later" pathway of medical applications. One is the well-known glucosamine-chondroitin combination, which was originally developed for joint repair and relief of joint pain in racehorses, then found its way into human sports medicine.

Our NMDA treatment also reversed the usual order of things. We had developed the NMDA-blocking treatment of compulsive disorder in pet animals first. That treatment then headed for human psychiatric use. We had "reversed the arrow"—using what we'd learned about animals to develop an effective treatment for humans.

While Dr. Jenike still uses off-label Namenda to help hundreds of OCD sufferers, the med is something of an orphan treatment for OCD. The patent we developed at Tufts was never taken up by a major pharmaceutical company. In June 2013, another medical research team discovered that the NMDA blocker, ketamine, was an effective treatment for OCD in people—exactly the same finding in our cribbing horse many years ago. Even though it is not yet used widely, it continues to show promise for treating OCD and various psychiatric disorders. We also discovered in the 1980s, that opioid antagonists caused penile relaxation in male horses and that they led to almost instantaneous evacuation of a horse's bowels. If we had capitalized on any of these discoveries, we could have hit the jackpot with the first Viagra-type drug and a drug to treat constipation. Now, some thirty years later, these treatments have become a reality.

Oh, well. Maybe in my next life I'll be a businessman!

I have long believed that OCD is fundamentally a single disorder that simply differs from species to species in the way it's expressed. To look deeper into animal compulsions, I needed to examine the genetics of affected animals, and perhaps also the structural changes in their brains. I believed that anything we found would help establish similarities between compulsive behaviors in animals and human OCD that could eventually shed more light on the human condition.

I contacted Dr. Edward Ginns, a pediatric neurologist and geneticist by trade and a former NIH branch chief. Ed had been hired by UMass to head up the Brudnick Neuropsychiatric Research Institute and is an expert at looking at the genetics of closed populations, such as the Amish demographic in Pennsylvania Dutch country. Pedigree dog breeds are, by definition, closed populations. We planned a study of certain compulsive behavior prone breeds, using the same approach that Ed had employed with human populations. We decided to study flank-sucking Doberman pinschers, tail-chasing bull terriers, cribbing horses, and two separate breeds of wool-sucking cats, Siamese and Birman.

We began by creating a DNA bank from affected and control animals to store for later analysis. I collected samples from dogs, cats, and horses, all of whom exhibited various animal compulsions. As scientists, we always require a control group for comparison, so I also took samples from animals who were not affected. We recorded detailed information about various symptoms the animals were showing, how serious the problem was, and what other problems they suffered from.

We chose to analyze dog DNA samples first because the canine genetic test was the only one then available—full genome maps of horses and cats were at that time still in development. And we processed the DNA from Dobermans first because the breed's compulsive behavior was manifest in the purest way. Flank sucking in the breed was clearly genetic, with up to 70 percent of some litters being affected. It can also be harmful to the dog, causing severe physical problems including lip calluses, buckteeth, and gastrointestinal issues, as well as intestinal obstructions that can be fatal. We held off analyzing the bull terriers, since that breed had seizure issues and a zinc deficiency problem—in addition to what appeared to be compulsive behavior—which could have clouded our conclusions.

In addition to flank-sucking behavior in Dobermans, we also observed that the affected dogs had a great penchant for collecting and sometimes arranging objects, a behavior Dobie owners called "shopping." One rescued Doberman was caught on camera arranging his collection of Beanie Baby toys in symmetrical patterns. On one day he would organize them in triangular patterns, the next in straight lines or cruciate designs. Most often the Beanie Babies were all the same type, all stuffed bears, for example. The Dobie usually laid each of his dolls facing the same way, up or down.

After his rescue and placement in his new home, it took this dog several weeks before he felt comfortable enough to allow his owner to give him a hug for the first time. The very next day, the Beanie Babies were arranged in pairs, one facing up and another above facing down as if they, too, were hugging.

After the painstaking process of collecting the DNA samples from the Dobermans, it came time to analyze them to see what was really underlying this clearly compulsive behavior. The results were exciting. We found an obvious glitch on chromosome 7, a clear discrepancy between affected and normal dogs. That particular region of the chromosome contained only one huge gene, known as neural cadherin, or CDH2.

The cadherin family of genes as a whole is interesting. Cadherins are implicated in numerous developmental disorders. This particular gene is expressed in the brain and is necessary for the proper formation of synapses, or the gaps between nerve cells. The nerve cells transmit chemical messages across these synapses, forming a chain of connection along which our thoughts travel, our internal organs are instructed and our actions are initiated. CDH2 is also responsible for the proper formation of NMDA receptors.

It all fit together. We were zeroing in on several factors that disrupted the natural flow of communication in the brain. This occurs on a cellular and even a molecular level. As with our former studies in PTSD research, we were going in deep. A series of minuscule neural miscues seemed to trigger the set of repetitive thoughts and behaviors that could be instrumental in driving OCD.

We took our findings to the National Institute of Mental Health (NIMH), whose scientists were intrigued. They agreed to look at CDH2 in DNA samples from people with OCD, investigating if the same brain glitch we found in Dobermans could be involved in causing OCD in humans. Regions of the CDH2 gene in people with OCD were found to be associated with an extreme form of OCD and provided genetic evidence of a known link between OCD and Tourette's syndrome. More recently, a South African study has shown that variations within the CDH2 gene are definitively associated with OCD in people.

Ed and I continued to work with our compulsive Dobermans, and joined forces with a canine geneticist, Dr. Mark Neff, of the Van Andel Institute in Michigan, to reanalyze our Doberman DNA samples. This time we used a more advanced genetic tool to flag other suspicious chromosomes.

Comparing seriously affected dogs with mildly affected ones, we made another exciting discovery. We found a region of great interest on chromosome 34, an area on the genetic chain that contained serotonin receptor genes. Recall that OCD responds to treatment with serotonin-enhancing drugs, such as Prozac. At one time, OCD was thought to be a serotonin-deficiency syndrome. Modifying serotonin levels in the brain is still the primary treatment strategy. Dr. Ginns and I reasoned that the anomaly on the CDH2 gene needed to be present for a dog to be susceptible to their compulsion. The *severity* of the disorder, however, was determined by the serotonin gene anomaly on chromosome 34.

My veterinary behavior resident at the time, Dr. Niwako Ogata, wondered if we could take a completely different approach to finding parallels between compulsive behavior in Dobermans and OCD in humans. She came up with the idea of looking at the detailed brain structure of affected Dobermans using voxel-based morphometry (VBM), a sophisticated version of magnetic resonance imaging (MRI).

The results yielded another eureka moment. The brains of affected Dobermans had structural differences very similar to those identified in humans with OCD, hoarders in particular. That revelation is particularly interesting, given that Dobermans are hoarders, too.

Our OCD work has had wide-ranging practical application. If you know the cause of something, it becomes a lot easier to develop logical solutions. For example, CDH2 interacts with proteins called catenins within the cells. The word *catenin* comes from the Latin *catena*, meaning "chain," and that's what we are looking at now, a chain or pathway activated by CDH2. We know that CDH2-catenin complex is involved in learning and memory and believe it is involved in propagating OCD.

It is no surprise to us that a catenin antagonist has recently been shown to help recovering alcoholics overcome their compulsion to drink. Fixes of this type take the kind of information we are seeking and finding, and though such research to address the mechanisms involves a whole lot of trial and error, the results, when they come, are well worth waiting for.

Ozzy the Oddball was a ten-month-old, 22-pound castrated male schnoodle, a schnauzer-poodle mix with a hangdog look on his large, pretty face. Sadly, ever since he was six months old Ozzy exhibited a number of odd behaviors. He would "stargaze," disconcertingly staring off into space at nothing, or engage in "fly snapping," trying to catch imaginary flies with his teeth. When not busy with those endeavors, he might obsessively lick either himself or various objects around the house. When we saw Ozzy at the clinic, he showed clear signs of compulsive behavior, perhaps complicated by a type of partial seizure.

With every animal we treat, we try to address issues with both pet and owner. Owners play a key role in helping their animal get better. We need the partnership of owners in administering medications, and helping with behavior modification. For Ozzy, we recommended that the owner withdraw attention when Ozzy engaged in his abnormal behaviors. That way, Ozzy would not be rewarded for behaviors the owner didn't want.

Ozzy also needed aerobic exercise, a regimen reinforced by a consistent daily schedule and a more interesting home environment. On the pharmacological side, we prescribed Namenda, our reliable NMDA blocker. Over the next few weeks, the owner rated Ozzy as 50–75 percent better.

But animals, just like human beings, are subject to accidents and reversals. After he started his therapy, Ozzy had a taxing experience in Florida. He became too frightened to cross a boardwalk. Soon afterward, his owners had to board him in a kennel for a time. The resulting stress, combined with a switch to a cheaper drug, dextromethorphan, led to a full-force return of his strange behaviors. He resumed stargazing at near full intensity and began obsessively rubbing his eyes and licking his paws.

What more could we do? We had seemed to help Ozzy briefly with his compulsions, but now we were almost back to square one. It would turn out that Ozzy and dogs like him were suffering from partial seizure activity expressing itself in the form of OCD-like signs. The link between partial seizures and compulsive behavior was a revelation first suggested to me by my earlier bull terrier work and later confirmed by dogs like Ozzy. It is unclear why OCD tends to occur with epilepsy—both in people and animals—and we'll explore that link further in the next chapter.

CHAPTER SIX

The Dog Who Was Afraid of Puddles

Autism, Epilepsy, and Rage

Fear is no more fun for dogs than it is for people.
—PATRICIA McCONNELL, PhD

Some animals are born with genetic tendencies to develop behavioral problems, and some animals have behaviors instilled into them by bad owners. The curse of an insensitive or preoccupied owner is terrible for any dog, but when an entire breed has genetic issues, even the best owner in the world can't help. That's where our work at Tufts comes in: we research the inherited behaviors of animals.

In the late 1980s, I found myself helping to run a dog-training class at the North Grafton Campus of Tufts University. Heading up the class was Brian Kilcommons, a dog trainer who was influenced by the legendary British expert Barbara Woodhouse and learned the core principles of his trade from master trainer the late Captain Arthur Haggerty. Out on a ball field on campus one day, he taught a group of dog owners how to properly control their pets on leash. At Brian's request, I took over a bull terrier, named Blizzard, from its owner in order to demonstrate how to grasp the leash, how to walk without the dog pulling, how to do left and right turns.

Blizzard was a one-year-old white male. I noticed that, whenever we halted and I spoke with the owner, Blizzard would begin to lope lazily in circles, chasing his tail. I didn't think much of it until the owner took me to one side at the end of the class. It turned out that he was a local bull terrier breeder, and he had a problem at his kennel. Several of his dogs chased their tails, just like Blizzard.

His story sounded familiar. By coincidence I had recently read a case report in the *Journal of the American Veterinary Medical Association* on tail chasing. The subject was also a white bull terrier, also a one-year-old male. The report might as well have been about Blizzard.

As the owner and I discussed the report, it suddenly hit me that the condition was most likely genetic, given that it had appeared in a dog almost identical to Blizzard. If only I could get my hands on a few more of these "bullies," it would be a fantastic opportunity to get to the bottom of what was going on in the breed.

Back then, the explanation for tail chasing was that it was a stereotypy, a pointless, mindless behavior. Compulsive disorders hadn't been suggested as a cause of these behaviors in dogs yet. Nonetheless, the tail chasing of bull terriers was different from that of other breeds. In bullies, it was associated with other bizarre behaviors, including strange phobias such as balking at rain puddles. It was also sometimes paired with a tendency toward explosive aggression. Because of these oddities, I was more inclined to think of tail chasing as a neurological, seizure-based condition, which few other scientists believed possible.

Two other cases that I came across in addition to Blizzard's strengthened my opinion. One particularly severe tail chaser halted his schizoid behavior for only two reasons: to sleep when totally exhausted and to grab occasional mouthfuls of food. He spun so intensely and for so long that he wore the pads off his back feet and had to have his paws bandaged to prevent further damage. This dog was another white bully.

A large, Boston-based veterinary hospital referred the dog to me after a week of costly and exhaustive neurological testing. He arrived at my home in North Grafton on a Sunday afternoon.

I immediately administered an injection of Valium in an attempt to allow the dog some momentary peace. An antianxiety treatment for people, Valium is also a powerful anticonvulsant, and I wanted to explore the possibility that seizures were causing the bully's distress. The drug is often used to treat *status epilepticus* in dogs, a very dangerous condition in which the brain experiences a constant state of seizure.

Within a few seconds of the injection, the bull terrier became totally calm, walking around my kitchen just like a normal dog. The sudden change in his behavior was close to miraculous. Unfortunately, Valium lasts only minutes in dogs and, twenty minutes later, the beleaguered guy began to lope around again in wide circles. Soon thereafter, he resumed his tail chasing at its previous frenetic level. The sad conclusion to this story is that the owners were so financially and emotionally drained that they could not pursue any of the treatment options I suggested, and the tormented dog was put down.

But this white bull terrier's initial positive response to an anticonvulsant medication suggested that we could find a treatment for tail-chasing dogs. My colleagues and I eventually demonstrated that seizures are involved in the "bull terrier syndrome," by performing EEG and CT scans of seven tail-chasing bull terriers and five unaffected "control" dogs. All seven tail-chasing bull terriers showed an epileptic brain-wave pattern. All but one of these dogs also had hydrocephalus, a structural abnormality often called "water on the brain." Five of the seven tail-chasing dogs were successfully treated with that magical anticonvulsant, phenobarbital. Yet some veterinarians ignored our findings and continued to describe all tail chasing either as a stereotypy or compulsive disorder.

To better understand the breed, I collected more information about bull terriers. I also became friends with an outstanding bull terrier breeder and bully enthusiast, Marilyn Drews, who kept me well supplied with cases and information. In the course of our discussions, Marilyn sent me *The Bull Terrier* by E. S. Montgomery, a classic book published almost seventy years ago, as well as a children's book called *Boodil, My Dog*.

The Bull Terrier is very hard to find, so it was a great treat to receive a copy. The book describes how bullies are derived from the bulldog, the English white terrier, and Dalmatians, as well as a few other breeds. Bull terriers are grouped as "white" or "colored," which includes fawn, brindle and white, or black-and-white coat colors. Bullies have one of the highest prey drives of any breed. One champion bull terrier was said to have killed hundreds of rats in a single hour at a "sporting" pit in London. Citing examples of the behavior even way back in the late 1800s, the book sounded an early ominous note, observing that some bull terriers chased their tails.

Surprisingly and in its own way, the children's book proved equally informative. The character of Boodil, a classic bull terrier, often acted the clown. She was obstinate and quirky, just like a real bull terrier. Boodil also spooked at puddles. She was afraid of the vacuum cleaner. And she froze underneath overhanging plants and bushes, a behavior known in the bull terrier world as "trancing."

I became determined to understand—and to help cure—the tail chasing of bullies. Together with Dr. Alice Moon-Fanelli, PhD, a behavioral geneticist, we collected detailed behavioral information about affected dogs and healthy control subjects. We also took blood samples from as many dogs as possible for later DNA examination.

All together, Alice and I painstakingly analyzed the behavioral traits of a total of 333 dogs, slightly fewer than half of whom were tail chasers. We found that males were more likely to be affected than females, and that tail-chasing dogs also frequently displayed explosive aggression and trancing.

And then it hit me: The bullies could have a canine version of autism! This idea certainly jibed with all the significant signs and associations that we were otherwise struggling to explain. Sometimes in science, all the data in the world can't prepare you for a eureka moment. You can't plan to have one. Instead, lifelong training and experience suddenly will kick in, and the answer reveals itself. My eureka moment was exactly that—a moment—but it had taken a lifetime's study of animals in order to get there.

Now we just had to prove it.

And before we could prove it, I needed to learn more about autism in humans.

Autistic children tend to have robotic movement disorders. Some rock back and forth. Others run in circles or flap their hands or spin in circles. Autism is more prevalent in boys, and it is associated with explosive aggression, trancelike staring behavior, repetitive movements, obsession with objects, and self-injurious behavior. Also, about a quarter of autistic people apparently suffer from full or partial seizures, which can take the form of unexplained staring spells, confusion, unexplained irritability, and aggressiveness.

Tail-chasing bull terriers also display these behaviors. Alice reminded me that many bull terrier owners described their dogs as "socially withdrawn." Some had even asked her if their dogs could be autistic. A study of a large number of tail chasers confirmed that their owners regarded them as asocial and that they showed significant preoccupation with objects. It seemed like our case was halfway to being proved.

Dr. Ed Ginns had extracted and cataloged bull terrier DNA from our dogs even though neither he nor we had the facilities to analyze the samples. Fortunately, we were able to interest Dr. Elaine Ostrander, a branch chief at the National Institutes of Health and a world-renowned canine geneticist, in running our samples on a state-of-the-art Illumina "chip," which maps gene sites on DNA and allows us to see where anomalies occur. It took months, but the results were worth waiting for: Affected bull terriers had a genetic anomaly on the X chromosome. That made sense, since with an X-linked problem, males can be more frequently affected than females. One well-known X-linked condition affecting people, for instance, Fragile X syndrome, is a genetic condition that causes intellectual disabilities. Fragile X is the most well-documented single-gene cause of autism.

Interestingly, people with Fragile X syndrome have large protruding ears, a long face, and a high-arched palate, as do bull terriers. Behavioral characteristics of Fragile X include movement disorders similar to those of bull terriers as well as atypical social development, particularly shyness and limited eye contact. Bullies seemed to have all the behavioral characteristics of Fragile X syndrome.

We really thought we were on to something, but further genetic testing proved our optimism to be premature. Reanalysis of the same data and whole genome genetic sequencing of two dogs—one severely affected and one not—failed to confirm our conclusions.

Such setbacks and dead ends crop up often in scientific research, especially in genetic testing. We just had to swallow hard and carry on, hoping eventually to discover the genetic cause of this obviously inherited condition. We went back to the drawing board and are currently reclassifying our dogs in degrees of severity and making sure that none of the control dogs had any hint of affliction, which may have interfered with the study.

In the meantime, a different way of studying the syndrome opened up to us. A medical expert at Tufts, Dr. Theoharides, who has been studying aspects of human autism for many years, was interested in what we'd discovered about bull terrier syndrome. Theo, as he is known, had demonstrated that autistic children have elevated serum levels of a peptide called neurotensin in their bloodstream. This was exactly the kind of biomarker we needed in order to make the connection between the human condition and the autism-like syndrome we'd found in the dogs.

We collected samples from affected and control bull terriers to measure neurotensin levels and sent them to Theo for analysis. A few weeks later he delivered the results: The average level in the affected bull terriers was significantly higher than the level in unaffected dogs. This could only mean one thing. Our tail-chasing dogs were almost certainly autistic.

News that Theo and our team at the veterinary school were studying a canine model of autism reached the Tufts University president, Dr. Anthony Monaco. In his research career, Dr. Monaco made his name working on the genetics of speech disorders and autism. Monaco and his team at Oxford University in England found that, in two families that they studied, glitches in a gene called cadherin-8 (CDH8) caused susceptibility to autism and learning disabilities.

We had found a different cadherin gene involved in our Dobermans with compulsive flank and blanket sucking, and a suggestion of involvement of yet another cadherin gene in the bull terriers, so we were particularly interested in his findings. After we finish detailed studies of these genes, we believe we can improve diagnosis of the canine condition and help develop new treatments for dogs and humans.

Dr. Theoharides and other experts have suggested that flavonoids may be helpful in controlling aberrant behavior of children with autism. These yellow crystalline plant phytonutrients are responsible for the rich color of fruits and vegetables. Flavonoids can act as "biological response modifiers." Luteolin in particular, a flavonoid found in leaves, rinds, barks, clover blossoms, and ragweed pollen, has been demonstrated to have antioxidant, antiallergic, and anti-inflammatory properties. We plan to create a bull terrier study to evaluate the efficacy of luteolin for treating affected dogs.

A bull terrier aficionada in Britain, Terry Heath, removes additives, artificial colorings, and preservatives that are mixed into many foods that we and our dogs consume. When these artificial substances are excluded from a tail-chasing bull terrier's diet, the condition improves markedly. Top psychiatrists agree that it can be helpful to remove these artificial substances from the diets of people with certain psychiatric disorders, including attention deficit hyperactivity disorder (ADHD) and possibly autism. So while autism may have genetic roots, environmental factors seem to play a role in the activation of autistic-like behaviors.

At the moment, however, the only available treatments we have are not cures but palliative drugs in the Prozac family, as well as NMDA-receptor blockers and anticonvulsants. The range of medications represents a motley crew of nonspecific, behavior-modifying therapies. They might help alleviate some behaviors but don't treat the root cause, because we don't know yet what that is.

If we were able to establish more clearly what is going on with tail-chasing bull terriers such as Blizzard, we could treat them more precisely and successfully. And any new therapies we develop might work in autistic children, too. Either way you cut it, it would be a clear win.

Seizures are often associated with autism. There are different kinds of seizures. They don't necessarily have to be of the full grand-mal type, which involve the entire brain, and during which dogs might convulse, vocalize and salivate. Partial seizures, as mentioned previously, affect only a certain region of the brain. Depending on the brain region affected, partial seizures in humans, dogs, and cats can have varied expressions. The results of partial seizure activity are always distressing for the patient. In pets, they can disturb the owners, too. When the partial seizure triggers aggression, owners can find themselves in danger of being attacked.

Partial seizures can take the form of staring off into space (trancing), attacks of rage that seemingly come out of nowhere, or other bizarre expressions.

Like bull terriers, German shepherds are seizure-prone. And as we saw with Lief in chapter 5, German shepherds are prone to tail chasing. It is not uncommon to have a German shepherd suddenly develop grand mal seizures at around one to two years of age. But if a seizure were confined to a certain part of the brain, its manifestation could be tail chasing. Because the shepherd's tail chasing is often classed as a compulsive behavior, it's treated using anti-compulsive medication such as Prozac. But tail-chasing German shepherds do not respond well to this line of therapy. Typically, the improvement is rated at 20 percent or less. What does seem to work in many of these cases is adding an anticonvulsant into the mix. Because partial seizures manifest themselves in so many different ways, and can originate in so many different parts of the brain (hence the term "partial"), recognizing and treating them is challenging. I've encountered dogs who variously get intense rage, experience nighttime aggression, show practically paralyzing fear, chase their tails, stare at the sky, smack their lips, snap at imaginary flies, and gulp in air. I found support for my treatment of partial seizures from a study of dogs with a bizarre EEG pattern that was improved by anticonvulsants. One of the dogs displayed fly-snapping behavior.

Tail chasing and fly snapping are probably triggered deep within the brain, in a region called the hypothalamus. In humans, the hypothalamus serves multiple functions, including the regulation of sexual behavior, appetite, and emotions. In dogs, this brain region facilitates the same functions, including the appetitive phase of predatory behavior. This is why partial seizure activity in the brain region likely causes predatory-type responses, such as tail chasing and fly snapping, in dogs. These happen more commonly in dogs with high prey drive, such as terriers and herding breeds. The reason may be simply that brains of those breeds are wired in a way that makes them susceptible to seizure activity of a predatory type.

Laboratory experiments with cats have demonstrated the involvement of the hypothalamus in predatory behavior. Researchers put rats into cages with cats who remained passive and did not spontaneously attack the rats. But when the scientists activated electrodes that had been implanted in the cats' hypothalamus, the previously mild-mannered cats suddenly turned predatory and attacked the rats.

The hypothalamus is only one of several parts of the brain where seizures can occur. A few years ago I got to know a white female bull terrier called Stella, who entered my office one morning as if she were walking on eggshells. Her tentative slow-motion gait resembled the dance step called moonwalking.

Stella's overall behavior revealed that she lived in abject fear. Her extreme fright over almost everything meant that she didn't have much of a life. She had become a recluse, would not play with the other dogs in the home and could not even leave the house. During the consultation, I accidentally dropped a manila file. Stella jumped almost out of her skin.

I had never seen such extreme fear and reactions in a dog. There was no apparent reason for it. Stella's history was unremarkable. She had been well socialized and had never been abused. Other environmental factors were discounted, one after another. Poor Stella. I felt so bad for this craven, frightened dog.

The usual explanations for fear didn't fit her condition. After running through and discarding possible diagnoses, I fixed once again on complex partial seizures. But this time I considered that perhaps Stella's seizures originated not in the hypothalamus, like the enraged or tail-chasing dogs, but rather in the amygdala, or the "fear center" of the brain. An ongoing aberrant electrical discharge in this region might present as extreme fear. I ran an EEG. Sure enough, the results showed abnormal epileptic-like activity. I immediately started Stella on a treatment with phenobarbital.

The outcome was nothing short of spectacular. Within days Stella was a different dog, running with the other dogs in the back garden, playing a game of "king of the hill" on a mound of dirt. She generally behaved like a normal canine. Her fear was gone. Stella got her life back or, to put it in terms of a recent movie title, got her groove back. She later became somewhat tolerant of the medication, which is common, and some of her fearful behavior returned. But she remained about 70 percent improved.

As I treated more dogs with symptoms of partial seizures, I found more, unexpected triggers for them. For instance, sunlight can set them off. Benny, a seven-year-old spayed female German shepherd cross, was afraid of sunlight as well as flashing lights. Benny's owner showed me a heartbreaking home movie of Benny on the back porch, glancing anxiously at the sun, shivering with fear, and salivating profusely.

Benny's history was complex. She had been hit by a car and afterward developed bouts of chewing objects, salivating, and hiccuping. The local vet diagnosed the chewing and hiccuping as trauma-induced seizures, and treated Benny with phenobarbital. The chewing and hiccuping decreased in frequency but did not entirely stop. I reckoned that Benny's fear of sunlight might be attributable to partial seizure activity, too.

But how could we treat her? A change in dose of the anticonvulsant, or the addition of other drugs, might make her life more bearable, so we tweaked the phenobarbital dose and added another anticonvulsant, potassium bromide, to her regimen. And, for what the owner termed "bad Benny days," when the sun was particularly bright, he could also give her a Valium-type drug, so that Benny had a little something extra to ease her troubled mind.

She responded well to the treatments. Eventually, Benny could tolerate light in all its forms much better than she had since being hit by the car.

Bernie, a beautiful Bernese mountain dog, was brought to me because he had begun to stargaze, fly-snap, and lick and smack his lips at the age of two. Like Stella, he was also incredibly fearful, forever bolting terrified from room to room, glancing continually over his shoulder.

All these behaviors have been reported in people suffering from partial seizures. An EEG showed an epileptic-like pattern in Bernie's brain waves. This time, though, instead of the old standby, phenobarbital, we went with an herb, huperzine, which has anticonvulsant properties. It practically eliminated the problem.

I had been searching for an alternative treatment because of the disturbing side effects of phenobarbital, which include extreme thirst, excessive urination, increased appetite, weight gain, and, worst of all, liver damage. I'd heard about huperzine, an extract of Chinese club moss (*Huperzia serrata*), through a friend and colleague, Dr. Steve Schachter, professor of neurology at Harvard's Beth Israel Hospital. An expert on epilepsy, Steve had learned that huperzine was being used in cases of Huntington's disease, a genetic disorder that is frequently associated with seizures. After he had found that an extract of the herb was effective at preventing seizures in mice, he suggested we see if it could help dogs. We don't often employ herbal extracts at Tufts, but with Steve's glowing report of its efficacy and safety in dogs, we were confident we would have a productive and safe trial.

In fact, Bernie was one of the first of a series of dogs we treated successfully with huperzine, and he enjoyed freedom from his seizures for about six months. But, sadly, that wasn't the end of his trials. Bernie began showing signs of arthritic pain, even though he was still a young dog. This is a fairly common problem in large breeds such as Bernese mountain dogs. His local veterinarian prescribed a painkiller, tramadol. Unfortunately, tramadol can bring on seizures, and Bernie's stargazing, fly snapping, and strange lip behavior immediately returned. To get him back under control we reverted to phenobarbital, which helped markedly reduce all the behaviors he'd had from the seizures.

Of course, we hope that our series of successes in reducing seizures with huperzine will inspire a pharmaceutical company to develop it into an approved remedy for human cases of epilepsy. It's always gratifying to be able to give peace to an animal who has been suffering from unpleasant, disabling seizures. We've really given the dogs their lives back, and enabled them to interact with the people who love them.

After I'd treated dozens of dogs who exhibited partial seizures, a couple brought their four-year-old, intact female spaniel to see me. They announced at the beginning of the appointment that the spaniel was a "glugger." That was a term that I had never heard used before, which made sense because the owners themselves had coined it. The dog would gulp and swallow over and over. She would also air-lick, happily ingest dust bunnies and other debris, and sometimes lick or chew the linoleum floor covering.

Out in the yard, the dog would scarf down grass and dirt, sometimes throwing it straight back up. Then, as if a light switch had been thrown, her behavior would return to normal. These bouts of aberrant behavior had begun when she was one and a half years old. They occurred regularly thereafter.

When I asked the couple how long the attacks lasted, one said "minutes" and the other "hours." It turns out they were both right. When the dog was in this behavioral warp, individual glugging bouts lasted only a couple of minutes. But each attack was linked together with others in clusters, so a collection of bouts could last several hours. As in people with partial seizures, the aftermath can leave an animal exhausted, unreactive, or even fearful and aggressive. This so-called postictal period is one of the hallmarks of partial seizures, and it is used to help confirm the diagnosis. But the spaniel never displayed any confusion or sleepiness after these events. For the little spaniel, the seizures at the root of her problem might well be in the lateral hypothalamus, which controls feeding impulses and eating behavior as well as predatory behavior. The owners preferred that we not run an EEG on their dog to find out for sure if the brain waves showed a pattern of seizures, but they were happy to try her on phenobarbital to see if that helped her. And in fact, the dog's glugging episodes faded over a few weeks of starting the phenobarbital and never recurred. A full two years later the dog was still on phenobarbital and was still glug free. And the phenobarbital had not sedated her, as it can some dogs, so that was not the reason for her improvement.

Another pair of owners whose dog also swallowed oddly told me that their dog was a "snoofer." When I showed them the video of the glugger, they said, "Yes, that's it!" I treated that dog with phenobarbital, too, with the same great results. Interestingly, the snoofer, though not usually aggressive, became so immediately following a snoofing episode. This aggression after the seizure-caused "snoofing" could have been the result of miniature electrical aftershocks radiating out from one part of the brain. Signals emanating from the lateral hypothalamus, which controls eating behavior, could have been affecting a different part, the medial hypothalamus, which is involved in propagating rage.

Many dogs who suffer from grand mal seizures show aggression after the seizures subside. After I published the case of the glugger and the snoofer in the *Journal of the American Veterinary Medical Association*, I was immediately inundated with scores of owners around the country who reported that their dogs, too, glugged or snoofed.

Some people with partial seizures at times swallow rapidly at the beginning of an attack, and may also smack or lick their lips and work their jaws as if chewing. They feel queasy and become light-headed. They do not necessarily become unconscious, as people who suffer grand mal seizures often do, but they may experience a blurred concept of reality, during which they may exhibit violent episodes of uncontrollable rage and destructive behavior, after minimal or no apparent provocation. This condition, known as episodic dyscontrol, occurs in dogs, too. Yet as in the glugging spaniel, some people may come out of mini-seizure attacks with no apparent confusion, sleepiness, or ugly mood.

Cats can be affected by partial seizures, too. They can go through the "rage" that dogs show and the "episodic dyscontrol" that people show. These cats can also have occasional, unpredictable bouts of severe aggression toward people, usually their owners.

One cat flew off the handle one night when his owner came into the house from the deck. The attack was extremely violent, and the poor owner had to escape and spend the night elsewhere until her beloved pet calmed down. On another occasion, the same cat wouldn't let the owner into the kitchen, even to let her feed him or refill the water bowl. The attacks were random and very violent. This behavior sounded to me like the result of partial seizures and so we treated him with an anti-convulsant. After spending a few days in our hospital, the cat became supercalm—and he was never again aggressive.

I have seen several similar cases since. My most recent angry cat was named Lillian. One day, out of the blue, she attacked her owner, Michael, while he was sitting on the couch. Lillian had always been mellow and affectionate, but her meltdown was so severe and long-lasting that Michael was afraid that he might have to put her down.

To avoid that sad end, I volunteered to keep Lillian in my wife's practice area, downstairs at our home, until a course of phenobarbital kicked in. To be honest, Lillian was quite a handful at the time. We had to treat her with great caution so that we were not bitten, and for a few days we would interact with her only when necessary. But once again, the treatment worked like a dream and Lillian was able to return to Michael as her old affectionate self. She continues to do well to this day on the anticonvulsant.

Another feline manifestation of partial seizures is known as feline hyperesthesia syndrome (FHS). FHS is a strange episodic condition in which a cat's pupils dilate, its skin ripples as if it's trying to shake off flies, and it grooms itself frenetically along the spine. Cats with FHS may also show scary levels of aggression when in this state. Some dash away from unseen enemies. Others bite their tails until they're bloody. They also seem to hallucinate, staring off into space and then ducking as if being dive-bombed.

Some vets have suggested that FHS may be the feline equivalent of schizophrenia and have proposed treatment with antipsychotic medication. Others dismiss it as nothing more than a manifestation of muscle pain or discomfort along the cat's spine. Yet others think of FHS as a feline compulsive disorder. Some cats do indeed respond well to anti-obsessional drugs such as Prozac, but Prozac has some anticonvulsant properties at certain doses, so the story may be more complicated than "simply" a compulsive disorder.

I believe complex partial seizure activity is the root cause of FHS. Cats with FHS respond well to classic anticonvulsant drugs. Years ago, a Siamese cat called Spike was brought to me because he was chasing and attacking his tail. Spike was already being treated by his local vet with phenobarbital for grand mal epilepsy, specifically because he was prone to severe convulsive seizures. The tail biting was the last vestige of the strange FHS condition, so I prescribed a Prozac-like antidepressant to try to alleviate that final symptom. Fortunately it did the trick, and Spike and his tail greatly improved.

People who have frontal lobe seizures can also have extreme and involuntary self-injurious behavior, like head banging, skin gouging, and self-biting. Children with seizures may obsessively groom their hair.

People in the middle of a partial seizure often have enlarged pupils and experience a feeling of dread and fear. They may run mindlessly. Some have "out-of-body experiences" and visions or hallucinations. Because the symptoms of human cases of partial seizures so closely resemble those of animals, they support my diagnoses of partial seizures in pets.

But for further validation, I turned once again to my neurologist friend, Dr. Steve Schachter, the epilepsy expert. We'd run a lot of EEGs on dogs, for which they had to be lightly anesthetized so they'd remain still enough for the test. I was a little concerned that the results of the test may have been skewed by the effects of the sedation. By contrast, human patients remain fully conscious during an EEG and are fitted with a skullcap that has electrodes implanted in it to sense the brain's electrical signals, which are then transmitted to a recorder. Special filters are used in the analysis to remove electrical signals that merely show the movement of muscles due to consciousness, leaving the brain-wave activity plain to see.

I wanted to try to get EEGs of my dog patients while they were awake, and invited Dr. Schachter to the veterinary school to see if we could pull it off with a dog or two. Our first patient was a tail-chasing bull terrier, who had shown some scary aggression toward anyone who approached him. Dr. Schachter, it turns out, is fearless. Steve did not hesitate to kneel down on the consulting room floor and apply various needle electrodes to the dog's scalp. The dog lunged and snapped at him, but Steve was unfazed.

Steve recorded this dog's EEG for several minutes and the report showed a reasonable tracing of activity. The results were not quite conclusive enough for us to make a definitive diagnosis, but at least we had made a start. More importantly, Steve still had all his fingers.

People with epilepsy often show abnormal brain-wave activity between seizures, "spikes" and "sharp waves." These small bursts of electrical activity cause the release of minuscule amounts of neurochemicals in whichever brain region they occur, which affect the patient in various ways. For example, if the mini-events occurred in a region of the brain's emotional center, the limbic system, the subject's moods can be affected in subtle, persistent ways, causing fear and anxiety. This spiking could account for the persistent fear I'd seen in some of my animal patients.

Dr. Schachter wrote a book, *Brainstorms: Epilepsy in Our Words,* which collects the experiences of many of his patients. Some patients report anger as a trigger for their seizures, which brings to mind the springer spaniels and other breeds prone to rage. In almost 10 percent of Dr. Schachter's cases, intense fear heralds the onset of a seizure, which recalls Stella and her moonwalking. Other human subjects are like Benny, the light-sensitive German shepherd cross, and find that flickering sunlight triggers an attack of spontaneous swallowing. Some spin in tight circles, as do tail-chasing bull terriers. In both humans and dogs, sleep can be a trigger for seizures. Dogs we have treated take a while to return to normal following their partial seizures and one of Dr. Schachter's patients reported that even twelve hours after a seizure she still had a hard time talking. Sometimes full recovery can take as long as four to five days.

Through our research over the years, we've been able to relieve the suffering of hundreds of pets. Today, when I'm confronted with a tail-chasing dog or cat, I feel myself to be on a lot firmer footing than I was in the days before we could give animals EEGs and logically employ anticonvulsant medications developed for people.

CHAPTER SEVEN

The Horse Who Went "Harumph"

Equine Tourette's Syndrome

A good rider can hear his horse speak to him. A great rider can hear his horse whisper.

—**AUTHOR UNKNOWN**

There's a little joke we veterinarians like to tell: What do you call a vet who treats only a single species? Answer: a physician.

Yet we also like to say that we veterinarians treat people as much as we treat animals. Because, as much as my work is to fix animals, it's an important part of any veterinarian's job to give relief to owners who are worried about their pets.

Emily "Mimi" Edwards called me from Connecticut to discuss possible treatments for her Arabian stallion, Migdol, a valuable show horse who had won a number of prestigious competitions. Unfortunately, when away from the show ring, this beautiful horse suffered bouts of repetitive spinning and squealing. Most upsetting, he bit through the skin of his legs and flanks until they were raw. Mimi was upset by her horse's distress.

I couldn't diagnose the problem over the phone and, at this early stage of the discussion, bringing Migdol to Tufts was impractical. So my colleague Lou Shuster and I agreed to travel to Mimi's stable in Connecticut.

As we drove, we wondered out loud about Migdol's symptoms. Clearly, Migdol had some kind of repetitive behavior disorder. But was it a stereotypy, the kind of mindless "stall vices" we had encountered in our previous studies? Or did the behavior have a medical cause?

If it was a simple stereotypy, it was possible that the underlying mechanics were the same as cribbing. But that didn't quite make sense, since the symptoms were so different. The behavior was repetitive and seemingly pointless, for sure, but it was also unlike any of the common stereotypies that I was aware of at the time. Fortunately, Lou and I had remembered to take along opioid antagonist medication, just in case Migdol's behavior was, indeed, some type of stall vice.

When we arrived, Mimi led us to a dimly lit barn, where, in the farthest, darkest, dankest corner, stood Migdol. The extremely handsome light bay Arabian stallion wore a wire muzzle with a series of vertical metal bars to prevent him from biting himself. He resembled an equine Hannibal Lecter in *The Silence of the Lambs*. For added protection, in case the muzzle slipped off and he attempted to injure himself, a thick cover that looked like an oversized raincoat was draped across his back and secured under his belly with ties.

Mimi gently lifted up a corner of Migdol's blanket to reveal his scars. They were an upsetting sight. Patches of hairless, exposed skin showed where he'd bitten himself. Healed and semi-healed teeth marks pockmarked his flank. As Mimi was showing us his injuries, Migdol became highly animated. He began prancing around the stall and kicking out with his hind legs, causing us all to retreat.

"He's having an attack right now," Mimi said.

Migdol started whirling around in tight circles, glancing and jabbing in frustrated attempts to reach his brisket and flank, kicking out with his rear legs. His hooves striking the wooden walls of the stall boomed like thunder. This whole fit was punctuated with snorts, head shaking, and occasional sniffing of the ground.

After a few minutes Migdol quieted down again. Lou and I had never seen such behavior and we hoped that one of the medications we'd brought with us could give him some relief.

Mimi led Migdol into a larger space, an indoor arena attached to the stable, which had an observation box for trainers and spectators. She gently held the stallion's head while I gave him an injection of saline into the jugular vein. Lou and I then headed to the observation box, clipboard and score sheet in hand to count the minutes that elapsed before any change of behavior occurred.

At first, Migdol ran and pranced around the periphery of the indoor arena, seemingly enjoying his newfound freedom. After a few minutes, however, he displayed another bout of whirling and kicking. Almost midstride, Migdol started to buck and simultaneously kick out with his hind legs, squealing and pivoting all the while. That first attack was over in less than a minute.

Migdol resumed trotting around and exploring. After a while he slowed to a walk and started to sniff the sawdust-covered floor of the arena. A pile of horse manure attracted his interest and he spent a while investigating it. Then he suddenly exploded into another intense bout of spinning and trying to bite his own flank.

Our score sheet soon became covered with notations. The first two hours featured forty or so of these attacks, of various intensities and durations. Some bouts were perfunctory, while others were alarmingly severe.

At the two-hour mark, we had Mimi catch Migdol by his head halter. This time Lou held him while I injected an opioid antagonist called nalmefene. The drug has qualities similar to the naloxone that we'd used on cribbing horses, except that its effects last longer.

After we gave him the injection, we released Migdol back into the open arena. For the next two hours, as the nalmefene coursed through his blood and into his brain, Migdol showed no bizarre behavior at all. It seemed as though the nalmefene was effective on Migdol's condition, just as it had helped stop other horses from cribbing. But though we had a successful treatment, we had no diagnosis. Since we still didn't know what precisely Migdol was suffering from, Mimi agreed to bring him to Tufts, so that we could study him further.

The following week, when Migdol arrived at Tufts, we discovered something new about him: Migdol hated thresholds, doorways, or any transition from one environment to another. Getting him out of the trailer was our first challenge. Mimi finally managed it, even though Migdol reared, balked, squealed, and kicked throughout the whole process. Then, as she led Migdol through an arch leading to a large foyer, the stallion went crazy again, rearing and bucking and spinning around. This happened every time he went through a new doorway on his way toward his temporary stall.

Once Migdol was in his new quarters, he dashed back and forth, bucking and twisting, snorting and kicking. He then was calm for a few minutes, momentarily sniffed the ground, and began the whole upsetting display over again. During these attacks, he also attempted to bite himself, clacking his large incisor teeth close to his brisket, chest, or flank. Some of these attempts were accompanied by a strange *harumphing* sound. The cycle was continual. Although we did not observe him through the night, we assume this went on then, too, as horses don't sleep through the night as some animals do.

As we watched him the first day, Mimi told us about Migdol's heritage. Both sire and dam had similar problems, although Migdol's sire settled down once he was castrated. Migdol's half brother also displayed the same behavior. Sadly, that horse had to be euthanized after he kicked a wall and broke the pastern bone in his foot. The bloodline heritage was a pretty clear marker that Migdol's condition had a genetic component.

Migdol had started striking out and trying to bite himself when he was eighteen months old. The problem fluctuated over the years. His attacks were less severe during seasons when he was extensively bred. They worsened when he was given a season's rest from breeding. Being able to breed seemed to quench his fiery behaviors, while sexual frustration appeared to fan the flames.

In the stall, between Migdol's attacks, we repeated the saline control and, after an appropriate amount of time, injected the nalmefene. After the drug injection, Migdol almost immediately settled down, wandering back and forth and picking up occasional pieces of hay to nibble. The calming effects of the medication lasted about four hours, after which Migdol gradually returned to his familiar distressing behaviors.

Now that we had replicated the results from our trip to Connecticut, we conducted a four-day "dose-ranging" study. We progressively increased doses of the medication to see how long each dose's helpful effects would last. The doses ranging from 100 to 800 milligrams of nalmefene reduced Migdol's flank-biting attempts linearly. In the control period, after the saline injection, he suffered eighty bouts per four-hour-observation period. The frequency of self-mutilation decreased with increasing doses of nalmefene, and was virtually abolished with the 800-milligram dose.

Virtually abolished, but not quite. The score at the 800-milligram dose might have been an actual zero if we hadn't subjected Migdol to an ultimate challenge. Toward the end of the four-hour test period, we led a mare in heat past his stall. This mare "flyby" caused Migdol to attempt four desultory flank bites. As soon as the mare was gone he resumed his tranquil state.

Though Lou and I were not the first to observe flank biting in Arabian stallions, we were the first to uncover a suspected mechanism. A prior textbook report suggested that it was specific to stallions of this breed. The condition was either propagated by endorphin release or—as we later would come to believe—excessive glutamate activity in his brain. Probably the most important neurotransmitter for normal brain function, glutamate is a nerve cell messenger and it affects nearly every part of the central nervous systems. But when glutamate keeps stimulating nerve cells over a long period of time, it causes damage, as prolonged overstimulation is toxic to nerve cells. So a high level of stress is damaging to the neurological system as well as to psychological well-being!

In any case, we had discovered the rudiments of a treatment that might one day help afflicted horses. But opioid blockers were too short-acting and too expensive to use as a daily medication, so we needed to offer other ways to modify Migdol's behavior. We advised Mimi to give Migdol as much freedom and exercise as possible. She should either castrate the stallion or use him more consistently at stud.

Castration was not viable, since Migdol was a valuable breeding stallion. The prospect that the horse's distressing condition might be passed on via breeding did not deter Mimi, though, in fairness, the hereditary nature of the condition has not been conclusively proven. Once the stallion returned to Connecticut, the measures Mimi was able to implement did seem to help. Migdol still engaged in the distressing behavior, though at a slightly reduced frequency and intensity.

A few months after we'd worked with Migdol, a horse owner from Michigan, Jo Anne Normile, sent along a letter and a video that told the story of her American quarter horse, Dan the Man. This beloved horse had engaged in flank biting to such an extent that he became emaciated, failed to thrive, and ultimately had to be put down.

None of the equine experts Jo Anne had contacted had been able to give her a straight answer about what had caused Dan's issues. Some said it was likely a result of a skin condition. Others suggested he had "sand colic," which occurs when a horse ingests sand as he grazes. Still others argued that stallions who behaved this way were simply sexually frustrated. A detailed postmortem performed at a veterinary school ruled out a medical condition—at least one for which any tests existed.

The videotape Jo Anne sent me began with Dan being borne into her waiting arms. You could see the look of wonderment and love in the new owner's eyes. She and the horse grew to be exceptionally close. Dan became like a son to her. Even after the poor horse died, Jo Anne still looked for answers to his torment. She had dedicated her life to finding a diagnosis and a cure for his condition, and she wanted to enlist our help at Tufts.

The video record displayed the wretchedness at its very beginning, when Dan was about nine months old, shortly after he was castrated. Initially, he was just bucking and kicking more frequently and at first Jo Anne thought he was trying to escape from biting flies. But the behavior quickly progressed to flank biting, head shaking, and snorting that Migdol also engaged in. When the video showed Dan twirling in circles, biting his brisket and flanks during a midwestern winter with snow falling, the biting-fly theory was effectively ruled out.

The local vet tried to help Dan but, though some of the treatments were temporarily successful, the problem always returned at full intensity. Finally, Dan's constant misery and tremendous weight loss precipitated Jo Anne's painful decision to have him euthanized.

Jo Anne, Lou, and I decided to create a survey reaching out to other horse owners who might have experienced similar issues with their own stock, in order to find out how widespread this condition was and learn more about it. The three of us sat down to sketch out a series of questions regarding the odd, repetitive, self-harming behaviors of Migdol and Dan the Man. Jo Anne sent the survey to horse owners across the country and, in a few months, received scores of responses.

When we collected the data, we found that, far from being limited to Arabians, the condition affects all breeds of horse. It strikes males more frequently than females, but can affect either sex. The attacks usually begin when horses are young, and the average age of onset was eighteen months. The survey also showed that some affected horses made bizarre noises while engaged in the behavior. *Harumph* was the description of the sound given by more than one owner. One outlier response reported a horse that first exhibited the behavior at thirty-two years of age, after an accident. The flank biting had started after the horse had reared up, flipped over, and concussed his head on a rock.

Castration had a palliative effect on some stallions. No other medical treatment was reported to be particularly effective.

Jo Anne, Lou, and I all read widely in the medical literature about self-mutilation and tic disorders in humans. Jo Anne was particularly intrigued by references to Tourette's syndrome, a condition in which people sometimes shout inappropriately and make sudden motions or gestures. Samuel Johnson, the famous intellectual and author of the first English dictionary, had Tourette's. Johnson would whoop and gesticulate on crossing thresholds. Only those who knew him took it in stride. She highlighted the many similarities between Tourette's syndrome and flank biting in horses.

Symptoms of Tourette's syndrome in humans and the parallels with the equine condition we found include:

- Occurs mostly in males, as does the equine syndrome;
- Responds to chemical castration, which is effectively the same as surgical castration of horses;
- Arises in children of around seven years of age, the equivalent of eighteen months in a horse;
- Causes about 10 percent of affected persons to vocalize inappropriately, similar to the harumphing sound we had noted;

- Can cause hemiballismus, a fancy name for large involuntary movements of the limbs, such as the sudden striking out with an arm or leg;
- May cause repetitive sniffing;
- May cause head and neck motor tics, consisting of sudden jerking movements.

The last three behaviors all occurred in our surveyed horses. Some Tourette's sufferers also reportedly have trouble crossing thresholds, the same quirk displayed by Migdol. People with Tourette's syndrome sometimes unintentionally act out aggressively. For instance, they may act as if they are going to punch someone before suddenly stopping short. Likewise, normally harmless affected horses can appear as though they're going to attack. Afflicted people are often found to have above-average intelligence, and that was often how owners described their horses in the survey responses.

Signs of Tourette's syndrome are exacerbated by stress, as well as sexual thoughts and connotations. Again, there were obvious equine parallels, this time to the stud activity (or lack thereof) and frustrated desire in horses such as Migdol. Clinical signs of the syndrome usually dissipate when affected people are engaged in an absorbing activity, just as horses tended to cease the behaviors while working or running.

With all these parallels, Lou, Jo Anne, and I were pretty sure we had found an animal equivalent of Tourette's syndrome. People with Tourette's frequently have coexistent obsessive compulsive disorders, so Lou and I thought that we might be able to treat horses like Migdol and Dan with the same drugs we'd used for treating compulsive behavior in other animals. We wanted to use a drug that damped down naturally occurring brain neurochemicals that we postulated were somehow being overproduced. The overproduction and release of naturally occurring neurochemicals in specific brain regions is also thought to be involved in propagating Tourette's syndrome. Abnormal opiate receptor function has been demonstrated in people with Tourette's, and some sufferers have shown at least a partial response to treatment with opioid antagonists.

It was time to reach across the species barrier. Lou and I sent the Tourette's Syndrome Association (TSA) a video of affected horses. We described the common features of human and equine Tourette's syndrome, though we didn't know how our comparison might be received. The response might very well have been one of disgust and anger. People are not horses, sirs!

We were instead pleasantly surprised. The TSA agreed to fund a research project on the equine condition. Here was a chance to help horses and perhaps shed light on a human condition that bedeviled and distressed many people. At Tufts, Lou and I prepared our "laboratory," a stall at the large animal hospital with an observation window. We affixed a camera to the wall that was connected to a television monitor and video recorder. We also cushioned the walls with two-inch-thick pads, in order to prevent injuries to the horses.

We tried several different medications on nine horses who we believed were suffering from the equine equivalent of Tourette's syndrome and found the improvement with each one extremely compelling. It mirrored what we knew about the response of human Touretters. Even apart from the videos and notational data, by the end of the experiment we had a striking visual demonstration of the extremity of the problem. Our pads had been completely trashed. The stuffing had been literally kicked out of them. The nine horses with equine Tourette's syndrome had hacked them to pieces in a matter of weeks.

One of our subjects actually kicked out the observation port while Lou and I were taking notes. The window had three-quarter-inch iron bars on the stall side and safety glass behind that, but it was no match for the power of a horse's kick. Perhaps this was a particularly vigorous example of patient feedback, but it certainly had us cowering under the assault.

During the study, a crew working with a national magazine came to make a film about other experiments we were conducting. It happened that several of the TV crew members suffered from Tourette's syndrome, including the director Scott Handler.

I told Scott that I'd noticed that when two Tourette's syndrome horses passed each other in the stable, they would pause and touch noses. This greeting seemed to calm both animals. They expressed interest in each other for a while, almost as if they knew they had the same affliction.

Scott told me that there is often a moment of recognition when two people who have Tourette's syndrome meet. So saying, he walked up to one of the study horses and put his face to the bars. The horse wandered over to him and breathed on him. For several long seconds, Scott and the horse both peacefully investigated each other, with not a tic or grunt to be seen or heard from either. It was a magical moment.

Migdol was one of the horses volunteered for our trial and it was good to see him again. He was still exhibiting symptoms, though their severity remained diminished. During the trial, Migdol responded well to three weeks of treatment with an antidepressant/anti-obsessional medication, showing a 50 percent decrease in his self-biting behavior.

Once back in Connecticut, Mimi kept Migdol on that medication for months, and his behavior continued to improve. Mimi removed his iron mask and took off his protective blanket. Migdol was free at last. He occasionally halfheartedly would glance at his flank, but that was the extent of it.

Unfortunately, the medication is very expensive. Eventually, Mimi stopped administering it and Migdol reverted to a modicum of flank biting, but he never did it as ferociously or repeatedly as when we had first met him. Perhaps this was because of the changes that Mimi had made to his lifestyle, or perhaps the long-term use of the antidepressant had caused some lasting changes in Migdol's brain. Antidepressants can act as a sort of neurological Miracle-Gro, causing the development of new nerve cells in the memory center, the hippocampus, which is thought to explain at least some of the results in laboratory animals as well as people.

Jo Anne was gratified to have helped to discover the source of her beloved horse's terrible affliction and to have aided in finding a treatment. As she had hoped, horses can now be treated for this disorder so that it doesn't progress to the kind of grievous self-harm, physical decline, torment, and death to which Dan succumbed. For the love of one horse, many horses can now be saved.

And for the love of another horse, Pepper Belle, many people with Tourette's have been helped. Tourette's syndrome can be especially hard on children. Just at the time when they're trying to be accepted in school and other social circles, along come tics and shouts and involuntary movements. So it was to our great good fortune when Willie Ferrero, the owner of Pepper Belle, a Standardbred mare, heard of our study through the equine community and volunteered to help us by enrolling "Pepper" in our study. (Pepper Belle hails from East Longmeadow, Massachusetts, so she was close enough to trailer into the veterinary school for study.) During her professional career, Pepper Belle had competed in harness races across the Northeast and was a well-established champion. Willie was devoted to Pepper Belle and not just because she was a winning horse. He had known Pepper's mother, Cool Pat, and like Jo Anne and Dan, had been present for Pepper's birth, after which he was completely smitten. He watched as the slippery, uncoordinated foal first tried to master the art of balancing on her spider-thin legs. He grew to become familiar with her likes, dislikes, and foibles and essentially raised the mare as though she were his own offspring.

Pepper Belle was trained as a harness racer, and she loved to run and to race. But as a young adult she started to show the troubling behavioral signs of flank biting and chasing her tail. Colic was ruled out by his regular vet, and Willie turned to me because Pepper Belle's symptoms had become so severe that it was difficult to calm her so that she would accept the halter before races. Getting her into the starter's stall on the track was a major project and she would sometimes refuse to enter the trailer—just as Migdol had been afraid of passing through entryways and portals.

For Pepper Belle, the antipsychotic medication Haldol and a couple of sedatives temporarily suppressed her symptoms. But none of these meds was compatible with her racing career, because they made her tired, sleepy, and disengaged. Sadly, Pepper Belle's time as a competitor came to a sudden halt, after she fractured a sesamoid bone and bowed a tendon around the cannon bone during a race. "Bowing a tendon" means that the tendon fibers are torn and heal in a way that gives a bowed appearance to the back of the lower leg.

Since Pepper Belle was out of the racing game, Willie had the bright idea that she could help children better understand schoolmates who had disabilities and differences such as Tourette's syndrome. If kids got the chance to spend time during school visits with a fantastic animal like Pepper, a proven winner who also happened to suffer from Tourette's syndrome, the kids could learn to develop empathy and compassion. They might understand that a condition that makes people different doesn't prevent them from finding something at which they can succeed. Reaching out to the Tourette's Syndrome Association, Willie offered Pepper Belle as a mascot for their national movement.

It might have seemed an odd idea, using a horse to publicize a human condition, but to their credit, the TSA immediately saw the benefit and adopted Pepper Belle as a logo and mascot. In the effort to render Tourette's syndrome less of a stigma, they were willing to think outside the box. The TSA established a program to use the horse in educational efforts in regional schools.

Pepper's first appearance was at Birchland Park Middle School in East Longmeadow, Massachusetts. We all attended: Willie, his son and fellow jockey "Big John" Ferrero, Mark Levine (a representative of TSA), Lou, and I. Scores of children had gathered outside in the playground and watched as Pepper Belle paraded around, resplendent in full tack. Big John, dressed in colorful racing silks, rode behind her in a sulky, as the buggies used in harness racing are called. The two of them did a few circuits around the school's running track and then the children were allowed to come forward to stroke Pepper Belle.

After they had spent some time with the horse, the children went into the school auditorium and Pepper Belle headed back into her trailer. Willie showed the children a video of Pepper thundering around the track, taking first place over and over again. The director of the TSA then gave a talk about how Tourette's syndrome affects people. To complete the connection, I showed a video of Pepper Belle at our veterinary school, whirling and ticking away in one of her worst moments.

The children were impressed. This was the same horse? How could the horse attacking her own flank be the same impressive champion they had so recently seen racing and then befriended on the school's track?

The message was intuitive and easy to grasp: Tourette's syndrome is not to be feared or mocked. It is merely a quirk of impulse control. It is not a demonic possession that affects only the weak in spirit. Instead, it represents one aspect of the enormous diversity of expression that characterizes both human and animal behavior. The children at the East Longmeadow school didn't need much convincing. They'd seen an animal who lived with Tourette's syndrome and was also a successful race horse.

Pepper Belle made quite a name for herself as the mascot of the

TSA. Driven by his love for Pepper and his almost religious calling to help her and other sufferers, Willie wrote a tale aimed at children, *Pepper Belle*. But around the time he finished his book, Willie told me that he was suffering from leukemia, a recurrence, and that the cancer had spread.

"It's in my chest now," he told me, matter-of-factly. Willie died shortly afterward, but his son Big John now runs his farm in East Longmeadow. And Pepper Belle is still alive and well there.

Since we first met Migdol at that stable in Connecticut, Lou and I had come a long way in our understanding of Tourette's syndrome in horses. We also had a deeper appreciation of the trials and tribulations of people living with Tourette's syndrome. When we first observed Migdol's strange behavior, however, we never dreamed for a moment that we could uncover a behavior best described as Equine Tourette's Syndrome. Although we still need final confirmation of the genetics involved before we declare it equivalent to the human condition, we think we know what to look for.

There is a known link between Tourette's and OCD, with about 60 percent of Touretters also having OCD. A small slice of DNA in the human genome, specifically a gene called SLITRK1, plays a role in the development of nerve cells and also expresses itself in areas of the brain relevant to Tourette's syndrome. Disruption of this gene was found in a small percentage of Tourette's syndrome sufferers. Along with SLITRK1, we believe that the same gene we saw at work in dogs and people with obsessive-compulsive disorder (CDH2) might also be involved in Tourette's syndrome.

Dr. Shuster and I would certainly like to examine these genes in horses afflicted with Tourette's. With that in mind, we may well call on Pepper Belle to step up one more time and provide a DNA sample— to help her own kind as well as shed more light on the origins of the human condition.

CHAPTER EIGHT

The Dog Who Hated Surprises

The Many Faces of Aggressive Behavior

Our prime purpose in this life is to help others. And if
you can't help them, at least don't hurt them.

—THE DALAI LAMA

The big dog lay on the floor in my office consulting room, as calm as could be. After a few minutes, he rolled onto his side, let out a huge sigh, and fell asleep. Once in a while he would open one eye when he heard another dog in the next room, but mostly Bailey seemed like the calmest dog in the world.

Sitting in two chairs in front of me, Jack and Sarah, Bailey's two humans, described a very different dog. They lived in a suburb of Boston, where Bailey had free rein in their large house. Jack was a minister, with a gentle manner, fly-away hair, and large spectacles. Sarah was kindly and quiet with a ready smile. But despite their calm and warm demeanors, they were clearly in deep distress about the antics of their mixed-breed rescue, Bailey.

The stories were alarming. Recently, while outside the parsonage for his morning walk, Bailey had attacked a poodle that was innocently passing by. Even though the poodle was not harmed, his owner had reported the incident to the police, and was talking about filing a lawsuit. This would be bad enough for any dog owner, but a lawsuit could financially ruin Jack's church.

Bailey had certain definite triggers. Oddly, blue jeans as well as anyone in his territory seemed to set him off. A few days before the poodle incident, Bailey and Jack were out for a walk when they had been surprised by a man turning a blind corner around a hedge. Bailey had lunged, teeth bared, and ripped the man's jeans. Fortunately, the man was happy with Jack's offer of a new pair of denims and did not take the matter any further.

Bailey also was puddling in the house and compelled to bark at anyone who walked past the parsonage's large front bay window. He had even shown aggression to friends who visited. Bailey was not providing the welcome that visitors to their parson for counsel or comfort would normally expect. While Jack and Sarah loved their dog, they had to face the fact that Bailey was likely to injure someone else. They had been seriously considering giving Bailey up for adoption, a kindly euphemism for what we all knew would likely mean euthanasia.

Throughout Jack's scary tale of Bailey's attacks, the dog lazed around my consulting room at Tufts. I'd convinced Jack to let Bailey out of his harness when they arrived. Jack had been loath to do so, and he was surprised that Bailey seemed so calm once free of restraint. The dog had wandered around the room a bit before taking a nap. Bailey was hardly the most extreme case of fear-aggressive behavior that I had ever seen, though his owners' stories clearly indicated that he was territorial and anxious. Away from his own territory, which he guarded too fiercely, Bailey was a sweetheart.

The dog's behavior almost certainly stemmed from a bad start in life. Fear-aggressive dogs reach the high point of their belligerence at about two and a half years of age, the peak of their physical and social maturity. Bailey was a rescue, so his early environment was unknown. I would take bets that the first three months of Bailey's life were far from ideal, with substandard socialization exacerbated by frightening experiences.

The first thing I told Jack and Sarah was that Bailey's barking, which they had complained happened whenever someone passed their house, might be something they'd have to live with. As the old adage goes, if you don't want a dog that barks, get a cat. Most dogs engage in some territorial alarm barking when people or other dogs pass the home. Mine certainly do, but the majority of dogs are perfectly well behaved, even affectionate, once people are welcomed into the home. I did suggest that they move the couch away from the window so that Bailey couldn't jump up and see everyone passing by. We also discussed denying him access to the front room and getting blinds or curtains to cover the windows. As another old saying goes, what the eye doesn't see, the heart doesn't grieve. Sometimes simple changes in logistics can really help.

A tired dog is a good dog, or at least a better behaved one, and from what Jack and Sarah told me, Bailey didn't get nearly enough exercise. Even though I'd rate more exercise as a basic "square one" change, it could still make a difference. Dogs generally need at least one hour of off-leash cardio exercise per day—that's right, at least one hour.

Jack and Sarah seemed surprised by this, but even though they described themselves as "hardly spring chickens," they thought they could manage. Clearly, Bailey had some issues about his territory, probably because he was anxious. There was plenty of work we could do to help Bailey, but first I figured that Jack and Sarah were the ones who needed the most help.

Dealing with stressed-out owners is a large part of what I do, and of course most people have plenty of other things going on in their lives besides their dog's problem behavior. But they have to prioritize what to deal with on their plates—and sometimes what's on their dog's plate. High-protein diets can cause elevated levels of aggression, especially territorial aggression such as Bailey was showing. So Jack and Sarah needed to change Bailey's diet.

Bailey needed only a limited, maintenance-level of protein. When I first saw him, he was ingesting about 30 percent protein in his regular dry food. My advice was to bring that down immediately—that very day—to around 18 percent. Protein doesn't cause the aggression, but it will fan the flames, so reducing the protein in his rations would defuse at least some of the bad behavior.

With these measures in place, Bailey's aggression waned. He never attacked another person. The threatened lawsuit did not materialize and Jack's parsonage was safe. Bailey had a solid home with the good reverend and his wife and would not be rehomed, or worse, euthanized.

But there's a coda to this story that points to the complex interactions between people and their pets. It turned out that at the time of Bailey's "acting out," Jack and Sarah had been having a terrible problem with a loved one, and that person had died in their home as a result of drug addiction. Bailey had been the first to discover the body, and the person had been wearing blue jeans.

Perhaps the pooch's fear-based aggression was catalyzed by the troubled atmosphere in the home and the untimely death, which affected Jack and Sarah. When members of the household have changes in mood and behavior, this often affects their pet. Trainers often suggest that an owner's anxiety is transmitted to his dog, and there is some scientific evidence to support that contention. The stress that people face has been found to increase levels of the stress hormone cortisol in the saliva of people and dogs. Other studies have shown that owners' emotional states can be mirrored in their dogs. And not just negative moods—dog owners experience increased levels of oxytocin, the "love hormone," when they look lovingly into their pup's adoring eyes, and simultaneously the hormone is elevated in their pet. Following the psychological trauma of his owners, Bailey may have become overprotective of his "pack," which could have caused him to behave in an unusually defensive way.

Aggression is not just a minor issue in animals. It is the number one behavioral problem in all species, including human beings. "Bad" dogs, dogs who bite, dogs who attack other dogs, dogs who tear into other animals or humans, are put down all the time. Learning how to deal with aggression is crucial for both veterinarians and for health professionals who treat people.

For all its negative associations, aggression is sometimes a reasonable course of action in a particular situation: Parents naturally protect their young, and we all defend ourselves whenever our personal safety is threatened. But in other instances, aggression can be an over-the-top response to a marginal threat, challenge, or minor annoyance. Sometimes aggression is truly aberrant and dysfunctional.

I remember a social worker who attempted to help a prison classroom full of gang members understand the social realities of aggression. He asked the gangbangers to make two lists. One list was all the things that they would kill for. That list was fairly long and included perceived slights such as being disrespected by a rival gang, if someone unintentionally damaged their vehicle, or even if a passerby happened to scuff a prized pair of sneakers. The other list was stuff for which they were willing to die. That list was shorter. They would die for their family and for fellow gang members. That was it.

The point the social worker made was that, for most people, these lists were very short. The vast majority of folks have just a couple items on their lists. They would die for their family and for their country, for example, and if pressed they would kill for their family and country. The normal inventory of acceptable aggression is extremely limited.

Well, it's pretty much the same for animals, who have evolved a host of behavioral and other signals to show when they don't mean to cause offense and don't wish to fight. And even the aggressive signals of many animals are meant to tell others to back off—the Halloween cat whose back and tail are up is scared himself and telling you to give him space. He doesn't necessarily want to attack—he wants to be left alone. Most dogs are the same way. After all they're been bred for thousands of years to be domesticated and not aggressive.

Even fully "functional" animals will aggress if provoked to act in self-defense, and to protect their young or beloved family members. It is true that all dogs may bite, but the things functional animals would probably die for, if put to the test, is pretty much the same as the list for people—just substitute "territory" for "country" and "mates, close relatives, and offspring" for family and we're on the same page. I don't think animals would regard dying in defense of something precious as a form of self-sacrifice, though. It's more that they are prepared to fight to the death when the chips are down.

It's dysfunctional dogs who will attack for what seem—to us—to be trivial reasons. They have a hair trigger and are often described by their owners or handlers as being like Dr. Jekyll and Mr. Hyde. With gang members, too, anxiety, mistrust, and insecurity underlie their dangerous reactions and outbursts to perceived provocations.

What was remarkable about the two lists that the gang members made is how disproportionate they were. When the social worker forced them to really confront the question, it turned out there really wasn't too much the gangbangers were willing to die for. They were able to see that their level of aggression was totally unnatural and out of balance.

When I set out as a veterinary behaviorist in the mid-1980s, I felt I had a firm grasp of certain behaviors, especially compulsive disorders, but I also knew what I didn't know. And, taking stock, I realized I didn't know nearly enough about aggression. Sure, I knew how to sedate an aggressive dog. But sedating dogs doesn't solve the problem when a dog is aggressive all the time. I had to bone up on this subject, so to speak, and quickly.

There is more than one type of aggression in humans and animals, and each type requires a different treatment. According to the most well-known classification system for aggression, described by Dr. Kenneth E. Moyer in the late 1960s, there are several distinct types of aggression. Today, even though clinicians think highly of Moyer's taxonomy of aggression, we've learned about new types that simply do not fall neatly into any one of the categories he describes. For example, Moyer did not recognize medically catalyzed aggression, which we now identify as pathological aggression. Nonetheless, it's important to understand the kinds of aggression that our pets can develop.

MALE AND INTERMALE AGGRESSION

Males of almost all species tend to be more aggressive than females and testosterone fuels much human aggression. Most violent human offenders are male. Intermale competitiveness is at the root of many aggressive incidents. Men with high testosterone levels are more likely to be arrested for violent offenses, to buy and sell stolen property, incur bad debts, and use weapons in a fight. Surprisingly, rapists do not have exceptionally high testosterone levels. But *violent* rapists do—that is, attackers who viciously batter their victims in the course of a sexual assault. Athletes who take testosterone or testosterone-like anabolic steroids to enhance their performance demonstrate increased aggression. Violence can be a side effect of doping.

Males are "masculinized" in the womb when a short burst of testosterone is released by their embryonic testes, which promotes structural and functional changes in the brain. At the onset of sexual maturity a tidal wave of testosterone activates full-blown male behaviors. Think of a dimmer switch. You crank it to increase a light's brightness. That's what testosterone does. When the switch is turned down, the light is not off, just dimmer. Maleness is like a constantly lit bulb, and testosterone activates the switch to increase its brightness. Testosterone encourages social dominance, competitiveness, and impulsiveness in male animals. Most dog bites to children are delivered by intact males. As the old saying goes, if there are children in the house, there should not be testicles on the dog. Castration of aggressive male dogs is highly recommended as part of any treatment. Castrated male dogs, cats, bovines, and horses exhibit less aggression because, without testosterone, the dimmer switch is turned down. There is not as much call for macho aggression when the hormonal lights are low!

So you see, a neutered male animal is not an "it." He's still a biological male. Though intermale aggression decreases following castration, castration is not an absolute fix. That said, neutered dogs and cats mount less, roam less, and are generally nicer to be around. They also are less susceptible to developing some cancers.

Castrating pets should be an open-and-shut case. It's a simple operation, but in odd cases it can become complicated.

A cat owner named Gary brought in Smokey, his absolutely gorgeous male seal point Himalayan because of incessant urine marking. Smokey was also aggressive toward the other cats in the home and exhibited a bizarre habit of pawing at a glass-fronted fire screen in the family room. At night he would roam the hallways like Marley's ghost, seemingly in search of something and caterwauling as he trod his lonely path.

Addressing the urine marking was Gary's first priority. He lamented spending $30,000 cleaning up or replacing urine-soaked carpets and drapery. Smokey also urinated on and ruined various appliances, which rusted and had to be replaced if not immediately discovered, disassembled, and cleaned after targeting. Though he dearly loved Smokey, Gary was at his wit's end.

"Is Smokey neutered?" I asked.

"Yes," Gary replied, "but it was a real struggle." His local vet had performed the operation, but Smokey only had one descended testicle. This was deftly removed, but the search for the inevitable second testicle was fruitless. The vet planned a second, more invasive surgery to search for the offending organ in Smokey's abdomen and reportedly found and removed it. Afterward, he performed a blood test for testosterone and the result came back ultralow, comfortably in the properly neutered range.

Smokey passed all the tests I gave him, too, in order to triple-check that he had been successfully neutered. He did not display an unneutered male's telltale facial appearance (he did not have protective cartilaginous plaques in the temple region of his head giving the head a wide appearance); he did not smell like a male (that acrid, ammoniacal, male cat pee odor); and he did not have pronounced barbs on his penis (unneutered males have obvious spiny protuberances on their penis).

Satisfied that he was indeed a successfully castrated male, I set about treating the urine marking with the standard treatment of Prozac, which is 90–100 percent effective for this purpose in an overwhelming majority of cats. Smokey improved considerably, though not quite as much as I'd expected.

A couple of months later, Gary called to tell me that Smokey had broken out with urine marking at a level at least as bad as before. I tried a different antianxiety medication, buspirone (trade name Buspar). Smokey improved again, but not for long. I then sequentially tried other more long-shot medications, but they met with limited or no success. Nothing I tried worked.

Finally, almost a year after our first consultation, Gary called me in tears and said he was afraid he would have to have Smokey put down. He couldn't stand it anymore. I pleaded with him to give me one more chance. I wanted definitively to rule out the possibility that despite his surgeries and low-testosterone blood level, Smokey might still have a retained testicle. As the ultimate test, I would employ a prohormone that would boost testosterone if any testicular tissue remained in his body.

The test was done and the results nearly knocked me off my chair. Testosterone levels after the challenge test were sky-high. Whatever the vet removed during the abdominal surgery, it was not the missing testicle. Perhaps a lymph node?

I queried Rob, a surgeon colleague of mine at Tufts.

"How good are you at tracking down missing testicles in cats?" I asked.

"There's never been a testicle I couldn't find," Rob replied with dignified solemnity.

A surgery was duly scheduled. As Smokey slept quietly under anesthesia, Rob searched for, discovered, and removed the lost testicle in less than three minutes. Smokey went home to recover. His improvement was gradual but steady. After two months, there was no more urine marking, no more aggression, and no more fire screen pawing. Smokey became the perfect cat and I learned not to trust testosterone levels. They fluctuate widely depending on the breeding season, which is why Smokey's aggression and other habits waxed and waned, until his hidden stash of the forbidden hormone was confiscated.

MATERNAL AGGRESSION AND FEAR AGGRESSION

Maternal aggression evolved as a perfectly normal response to an outside threat against offspring. Mothers defend their young even to the point of being prepared to kill or die for them. At birth, moms of all mammalian species release prolactin, a pituitary hormone that promotes milk letdown and lactation.

But prolactin also facilitates maternal aggression. And the level of prolactin in the bloodstream after parturition exactly parallels the rise and fall of maternal aggression. Accordingly, one should be extremely careful not to appear to threaten the young of any mammalian species, especially when the mom is lactating. People who work around animals with young, whether pigs, cattle, horses, dogs, or grizzly bears, understand this. Unless they are really familiar with a mother and her brood, they stay as far away as possible and protect themselves against a sudden attack.

Prolactin is also found in male animals, having evolved partly to facilitate paternal care of the young. This is especially pronounced in species in which males have parenting responsibilities, including defense of their young. Perhaps women who want to know if a man will make a good father should have a prospective spouse's prolactin levels checked before proceeding further!

Although testosterone and prolactin can play a role in aggression, another important factor is how animals—and humans—are raised. Social isolation, mistreatment, lack of proper nurture, and other environmental disadvantages damage young creatures. A deprived upbringing and adverse events occurring in the crucial early weeks of life can be terribly destructive and contribute to a defensive type of behavior in dogs and cats that is commonly referred to as "fear aggression."

An incarcerated or abused puppy will often later manifest its mistrust as fear aggression, behavior that evolved to drive away the "scary monster." Dogs in the grip of fear aggression can be intimidating by growling, barking, and lunging. If these actions don't drive away the perceived threat, the dog may bite. At times, fear-crazed dogs will skip the growling and barking and attack right off the bat with teeth bared.

One such dog chose a very surprising target: me! I was walking my dog Rusty down a long curving path by the side of Lake Chauncy in Westborough, Massachusetts. I was heaving tennis balls ahead of us with a Chuckit! device, and Rusty was fetching and returning them. At one point he got too far in front, almost out of sight around a bend, so I called him to return.

To my surprise Rusty came charging back accompanied by another dog. The interloper hurtled toward me trailing its lead.

I was not quite sure I liked the look in the newcomer's eye, so I made what I thought were soothing sounds. "Good boys! What are you doing? Want to chase the ball?"

I let fly with the Chuckit!, hurling the tennis ball away down the road. Rusty did a U-turn and went after the ball, but the other dog kept coming. He launched himself at me and bit me in the leg, ripping my pants. Then he stood a distance from me and kept growling ferociously. I saw his distraught owner, a middle-aged woman, half-running, half-shuffling from around the bend toward me.

"Oh, dear," she said, as she approached. "I am so sorry. What happened?"

"He bit me in the leg," I said calmly.

"I've only just adopted him! This is the first time he has been free to run. I've always kept him on leash, but this time I lost my grip."

"No worries," I said. It wasn't a bad bite, I told her, barely a scratch, but he had ripped my pants.

"Can I buy you new ones?" she pleaded.

"No, no," I responded. "These pants were on their last legs anyway." My witticism earned a faint chuckle from the woman. Then she stopped and stared at me.

"You're not Dr. Dodman, are you?"

"Why, yes I am," I replied.

"I'm so ashamed! My dog bit Dr. Dodman, the behaviorist! How can you ever forgive me? I've read all your books and seen you on TV, and now this . . ."

I managed to convince her that I was not at all worried or offended. We actually had a brief "pets on the couch" conversation about canine behavior, how some dogs see certain people as scary, especially men, and especially those wearing specific clothes (like blue jeans or certain headgear) or carrying odd things (like a Chuckit!).

I walked home, the flap in my pants fluttering in the breeze. On the way, I meditated on the possibility that some dogs might maintain internal wish lists. "Bite mailman," could be one, and "bite animal behaviorist" could well be another.

Cats can show fear aggression, too, with a set of behaviors we behaviorists call "feline affective defense display." If they feel their territory has been encroached upon, or perceive some sort of outside threat, they will hiss, spit, snarl with open mouths, unsheathe their claws, and bite. They will also arch their backs and make their tail huge and puffy, like the Halloween cat in the popular holiday silhouette. This "piloerection" enlarges their profiles for maximum threatening effect. It is best not to approach cats who look like this! Wait until Halloween is over.

Preventing fear aggression is preferable to curing it. Actually, a totally effective cure is pretty much impossible to achieve, because psychological and pharmacological treatments have only limited success. To prevent it, you optimize a young animal's life in the critical period of development, which is between three and twelve weeks in dogs, and between two and seven weeks in cats. Proper socialization is crucial during these brief windows of time. Animals should be protected as far as possible from any scary encounters or other negative experiences, from which they "learn" to be afraid. They need to be actively socialized with frequent exposure to all the living things they are likely to encounter in later life, the encounters conducted under the most pleasant, warm, and friendly circumstances.

Veterinarians used to advise new puppy owners to keep their dogs isolated until vaccination was complete at fourteen weeks of age. Though this was recommended so that common viral illnesses and diseases could be avoided, it just so happens to be the worst advice for fostering a well-behaved dog. The first three to four months of life are when a pup is most malleable, in a behavioral sense. Completely isolating it from people and the world at large is likely to lead to unwarranted mistrust of unfamiliar people, other dogs, or even some inanimate stimuli. Breeders and owners should make every effort to socialize pups and kittens safely, while still avoiding exposure to sick and unvaccinated animals.

Once fear aggression does rear its ugly head, you have a few options, which I used with Bailey, the aggressive dog owned by the parson. Increase exercise and alter the environment to reduce the opportunities for the dog to react. Try a low-protein diet for the dog. Use clear, one-word commands, proper leadership, and firm control.

Human babies, too, have a critical period of development. Children need to be properly socialized and shielded from adverse early experiences. If repeatedly traumatized or neglected, children may grow up angry, fearful, aggressive, and unable to trust. Confident animals and well-adjusted people are rarely aggressive. They have no need to be.

MISDIRECTED AGGRESSION

As we have learned repeatedly from the news, aggression is not always logically directed and innocent bystanders can be affected. Purposely redirected aggression, sometimes termed irritable aggression, may also occur. A prime example is two dogs who converge at the front door when a visitor is outside. In this situation, both dogs are highly aroused, and the primary object of their aggression is unattainable. One of the dogs, usually the most fearful one, turns and attacks the other in what can be described as a "cheap shot" attack.

When a cat sitting on a windowsill sees something scary outside, it might, in its aroused and aggressive state, attack another cat in the household or even its owner. One cat I treated had become enraged when it saw a baby for the first time. Though the baby was quickly whisked away, the cat then focused its attention on his owner, pinning her in the kitchen for several hours. The woman had to keep the enraged kitty at bay with a broom.

I once saw this redirected aggression exhibited by my own cats, Cinder and her daughter Monkey. The two were always thick as thieves, bonding and spending every moment of the day and night by each other's side. They groomed each other, ate together, and curled up together like two inverted commas.

But on the summer night in question, a feline visitor came knocking on the screen door at the rear of the house. I heard some scary cat sounds, hissing, yowling, and spitting. I went into the dining room area to see what the fuss was all about and found Cinder and Monkey, backs arched and all puffed. By now the sounds were bloodcurdling. Their unwelcome visitor, which looked like an intact tomcat, had his nose pressed to the screen.

Suddenly, Monkey and Cinder tore into each other. It happened so fast that I could not see who attacked first. They were in full battle mode and it appeared as though they were going to get into a major league brawl. And these were two feline BFFs! Redirected aggression, collateral aggression, irritable aggression—a lot of aggression was in play.

I took one of the dining room chairs and, like a lion tamer, managed to back Monkey away from Cinder and then back into another room. Then I shut them off in separate rooms. This is what I had always preached to other owners of cats who acted out redirected aggression: harm would be limited as long as separation was immediate.

I kept the cats isolated overnight. It was with some trepidation the next day that I opened the door to Monkey's makeshift confinement. But the daughter came out and nosed a good morning to her mom as if nothing had happened the night before. Great Scott, I thought, the strategy turns out to be correct. Now I teach that lesson with greater confidence and more understanding.

Another common scenario in which redirected aggression occurs is when a well-meaning owner, and one who would not normally be the object of aggression, tries to separate two fighting dogs. The effect is like putting one's hands into a switched-on food blender. Apparently, even Queen Elizabeth II was once subjected to such aggression when she tried to separate dueling corgis.

In my time researching animal behavior, I have encountered unusual cases of redirected aggression. A farm dog I knew got shocked periodically by an electric fence. The jolt stirred up nascent aggression in the dog and he'd run some distance back to the farmhouse to attack the other dog in the household, who was often sleeping peacefully on a bed by the hearth. Another farm dog also redirected his aggression in an odd way: whenever someone approached down the long driveway to the residence, he would run to attack a nearby horse.

Irritable or redirected aggression also occurs in people. Think of an angry man punching a wall. The target of his aggression is not the source of his frustration. Sometimes irritable aggression *is* directed toward the source of the frustration. A person being physically restrained may strike out at the people restraining him, kicking, flailing, or even biting them. Exactly the same may occur in animals, for the same reasons.

People and animals showing this type of aggression tend to be more volatile than their more equable peers, which implies that the aggression is something of a personality trait rather than a chemical imbalance. For humans, a course in anger management is sometimes prescribed. In animals, the treatment is similar to that for fear aggression: increase cardio exercise, ensure an enriched and protected environment, proper nutrition that if necessary lowers protein intake, and sometimes mood-stabilizing medications. All help to level out mood swings.

INSTRUMENTAL AGGRESSION

Not all types of aggression are as seemingly weird or random as redirected aggression. Some forms of aggression help one animal to secure a selfish end at another animal's expense, and thus are termed *instrumental*. The aggression is not an end in and of itself, but rather occurs incidentally en route to getting something. Human examples of this type of aggression are when a robber punches a store clerk to gain access to the cash till, or when a child injures another child while stealing something that he wants. When someone is hurt or injured by the action, this is a by-product of the aggression, not its point.

Some veterinarians are skeptical that animals act out of instrumental aggression, but I believe they do. Sometimes, especially willful dogs will steal something right in front of their owners, and they might bite if the owners try to retrieve it. They may also dictate whether or not they will allow a person near their food bowl. They decide, not the owners, when they want to be petted and when the petting should stop. It appears to me that a dog like this is using aggression as a means to an end—that is, instrumentally.

With pets it is important to avoid situations that lead to conflict and to demonstrate clear control so that this kind of aggression does not continue. One way is to elevate your leadership status in the eyes of your pet. You can do this by having the animal *earn* valued resources by first obeying a command. This program has been variously termed learn to earn, nothing in life is free, or no free lunch. You get the point. So do the pets.

Instrumental aggression can also occur between cohabiting dogs.

Inter-dog housemate aggression is a treatable behavior. Because "inter-dog housemate aggression" is a mouthful, I choose to call it sibling rivalry, even though it may refer to two or more dogs in a house who are not actually related. Sibling rivalry usually occurs in the owner's presence. When left alone, the dogs often sort things out among themselves.

Problems can escalate if owners don't handle inter-dog conflicts correctly. Humans might inadvertently favor the underdog ("Oh, you poor thing!"), which snubs or shortchanges the true leader. It's easy to do when a cute though pushy young pup is around. Supporting the wrong dog as "number one"—or the favorite—disrupts the hierarchy of the pack. The underdog then forms an "alliance" with the supportive owner and is only brave enough to cause trouble when the owner is around to protect him.

When two dogs run to greet an owner, the race is on. Who will be greeted first? It's all about winners and losers. The dog that loses out should be the lower-ranking animal, but the "underdog" may attack and bite the other dog to get its own way. In this situation, owners often do exactly the wrong thing. Because they consider such a punishment to be logical and fair, the owners separate the dogs and give them both a time out. This impartial approach doesn't work with dogs. Separation of the dogs emboldens the lower-ranking dog. There are no real repercussions for his assault. The top dog doesn't get to make the point about his *right* to be first.

Dogs are not like us in respect to hierarchy. They don't need to be number one, but they do need to know where they stand. The true top dog within the pack should always be supported as the alpha. Owners who sanction the pack's naturally appropriate order can avoid sibling rivalry, ensuring that peace reigns between the dogs.

Just as in a wolf pack, in most cases the top dog is the older one. There comes a time when an aged or infirm dog can no longer defend his top-dog ranking, and at this point, he has to be protected from continual challenges from physically fitter, power-hungry peers.

Under normal circumstances, however, the older top dog should be first in everything, including being greeted first on the owner's return. To this end, the lower-ranking dog can be sequestered before the owners leave the house. In addition, the top dog should be fed first, petted first, given treats first, allowed priority access to high places, such as the owner's bed or sofa. The alpha should be allowed outside first, and should be allowed to ride farther forward in the car.

If there is a fight, the two dogs should be separated, of course. But the top dog should be petted after the fight while the lower-ranking dog is tethered or crated in the same room and forced to watch. This sends a powerful message about which dog actually is top dog. The owners, who are the true top dogs, support his position. Many people have a hard time with this process. It goes against our human sense of fair play. But without these measures dogs will continue to fight for supremacy.

TERRITORIAL AGGRESSION

Cats are not driven by hierarchy, rank order, and privilege as dogs and humans are. But housemate cats still sometimes get into serious tiffs with each other. There are several reasons why this happens. Cats are very territorial, so one of the most common issues that arises is a turf war. This type of inter-cat aggression—territorial aggression—can occur when a new cat is brought into the home. The treatment is separation and gradual reintroduction to allow the animals to slowly acclimate to each other. Sometimes the personalities of the feuding cats are so diametrically opposed that continued separation is the only solution.

Just as territoriality causes people to fight over land, territoriality in our pets explains many quirks of their behavior. Some dogs refuse to let other dogs anywhere in or near their yard, and some cats do not readily accept another cat into their home. Territory is where all creatures harbor their resources, their mates, and their progeny. Ethologically, territory is defined as an area that an animal will protect against intrusion by its *own* species. But dog and human lives are so entwined that dogs sometimes see both human strangers and other dogs as territorial invaders.

In South America, an advocate for the poor had to help establish the property lines in the crowded favelas and sprawling slums, where they had not been clearly established by authorities. How to decide where one family's yard began and another ended? The distinction was vital in order to elevate the poor to the status of property-owning citizens. The advocate's genius method of determining a property line was that a family's land boundaries began where its dogs started barking at strangers.

A dog's territory is not simply the house in which it lives. It is also the yard and the surrounding street frontage, areas that the dog frequently marks with urine. It often includes the car or truck in which it rides. I call the family vehicle the dog's mobile territory.

Dogs are definitely territorial animals. Unless you take certain steps, your dog, not you, will decide who can and who cannot enter the home. With kingly hauteur, they will allow you to stay around in their territory simply because you feed them. To control a confident dog's territorial aggression, owners need to be strong leaders and take care that no visitors are put at risk.

I once treated a hefty, fearless bulldog, Tank, whom I nicknamed Territorial Tank. One day, a friend of Territorial Tank's owner found the front door open a crack and let himself in. Bad idea. Tank's owner was in the shower. Tank was eating breakfast. Sensing territorial invasion, Tank left his meal and attacked the intruder. The friend got stitched up in a nearby hospital. Territorial Tank's aggression arose from his confidence and ownership of his home space.

A different version of territorial aggression is linked to anxiety and uncertainty. Such dogs might be better classed as "fear biters," but they only have enough confidence to come forward and attack perceived foes from the security of their own home, where they have the home field advantage. Treatment for such fear-based territorial aggression involves the same type of management as described above for fear aggression. You the owner need to assert unfailing control in such cases, especially at the front door but also inside the home when visitors are around.

A case of fear or anxiety-based territorial aggression I became involved with as an expert witness for the defense (yes, there was a lawsuit) concerned a pitbull terrier. A lining up of bad planets led to a whole load of trouble for all parties involved. The victim, a UPS delivery man, had all the traits of a person that an anxious dog might find concerning. The man was wearing a uniform; was six feet five inches tall; was frightened of dogs (particularly pitbulls); and started to move awkwardly, almost robotically, when he realized the dog was behind a desk at the place he was delivering the package. He hastened to leave the room but the dog pursued him and bit him in the leg. The result: a one-centimeter superficial skin-wound on the inside of one of his knees. The UPS man now claimed he had PTSD, couldn't sleep, had to have numerous psychological and psychiatric consultations; and could not perform his work functions properly (he sat in the van and honked the horn waiting for people to come out and pick up packages). The lawsuit against the dog owners was for $350,000 to make up for lost salary, expenses, and suffering. I went to bat for the dog owners, whose dog had never done anything like this before and whom they had no reason to mistrust. It was just a bad moon on the rise for them. The case settled for less—$160,000 I seem to remember—a fair outcome for the unsuspecting owners and unsuspecting victim.

Our front doors and foyers are so often battlegrounds. Progressive, step-by-step training of dogs is helpful to demilitarize the situation. Owners can and should desensitize their pet to feared or mistrusted strangers. Desensitization is a systematic approach as usual, in this case to front door introductions. Slow and steady does the trick. Neither the dog nor the visitor is pressed beyond the limits of their acceptance of each other.

For another useful strategy, counterconditioning, calls for arming strangers with high-value treats. These might be superfun toys reserved exclusively for times when people are visiting. A hot dog—the canine equivalent of an olive branch—also goes a long way toward cementing a friendship. Uniformed visitors, whether workmen, postmen, or UPS delivery people could during the course of their travels equip themselves with dog biscuits when visiting regular customers. The dog learns that these visitors aren't a threat. Rather, they come bearing gifts. A good mantra to bear in mind is that guests should "arrive as strangers and leave as friends." Alternatively, dogs can be secured away from such visitors to ensure that accidents do not occur. It's just not worth the risk with infrequent visitors, and, as you hear, the stakes are high.

If all else fails, antianxiety medication may help the behavioral methods succeed. One unusual and highly successful accidental treatment of territorial aggression in a dog occurred when the dog got hold of his owner's supply of medical marijuana and scoffed the lot. Apparently, the dog welcomed all visitors that day with a virtual "high four" and became a veritable love bug. It was flower power from the sixties all over again. Some avant-garde companies are now making a form of medical marijuana for pets. One of these preparations that we have tried—with some success—to treat various anxious conditions in dogs contains mainly cannabidiol (CBD), one of the many active cannabinoids in cannabis. CBD is also an effective anticonvulsant, so I feel sure cannabinoids like CBD will soon find their way into mainstream veterinary behavioral practice. Peace and love, everyone!

CHAPTER NINE

Two Spaniels and a Baby

Predation and Pharmacological Fixes

Humanity's true moral test . . . consists of its attitudes
towards those who are at its mercy: animals.

—**MILAN KUNDERA**

The woman at the other end of the phone line sounded as if she were at the end of her rope. "My name is Robin," she said in a shaky voice. "My local vet said you might be able to help. My dog is trying to go after my new baby, and to tell you the truth, I'm very scared."

The situation, and the woman's desperate tone, prompted me to take her very seriously. "Tell me exactly what's going on."

When she had brought her newborn back from the hospital a few days before, she had sat on the couch cradling her new darling. Her husband, Barry, brought in the couple's two English springer spaniels to meet the new addition to the family.

"Samson, the male, wriggled with excitement and ran across to see me," Robin recalled. "Then he began whimpering and whining and got into a frenzy. There was no controlling him. He started to bark at the little bundle and mouth at the bedding. Barry had to take Samson outside."

"Has the dog displayed more signs of unruly and unsettling behavior toward the baby?"

"He's definitely not getting better," Robin replied. "We tried putting him in the basement, and he just barked and barked until the neighbors complained. When we put him outside, he actually attacked the vinyl siding of the house, and tore through a screen door trying to get inside."

I told her it was vitally important to keep pets and baby separate and instructed her to bring the dogs in to see me the next day.

Samson and his female housemate, Delilah (of course), were beautiful, normally obedient dogs. But during our consultation, they told me tales of Samson's predatory impulses, such as the time he snapped up a friend's parakeet, not releasing the poor bird until Barry pried open the dog's jaws.

While at my office, Samson and Delilah were the picture of sweetness. But I wanted to see how the dogs behaved in their home environment, so I visited the residence the next day. There I witnessed what I considered a dangerous situation. When the baby cried, the dogs became agitated. Delilah ran to the bedroom where the crib was, put her paws up on the side, and began weaving and jumping as if searching for something. When Robin placed the baby down, both dogs started to whine and wriggle, pushing themselves against the crib. They were on a mission. Samson's whining rose to a crescendo.

"You have to keep the baby and the dogs completely apart," I said. The male dog was the primary offender in the situation, and Delilah was simply piggybacking on his aggression. So I counseled thirty minutes of aerobic exercise every day for the dogs and a change in diet to a strict, nonperformance ration. Even though both dogs were well trained, I ordered a crash refresher course of obedience sessions.

"We will work on a program of gradual introductions between Samson and the baby. We'll reward him for remaining calm and direct him when he begins to lose control. But, outside of these periods, separate the dogs and baby."

During my initial visit, we conducted a training session designed to acclimate the dogs to the presence of a new addition to the household pack. I waited in the bedroom where the baby was in her crib and Robin brought both dogs in on a leash. They were agitated and whining. I instructed Robin to give them commands such as, "Sit" and "Lie down." Samson and Delilah both obeyed but continued to whine.

"What do I do now?" Robin asked.

"Just take the dogs out of the bedroom and get them under control in the living room."

We regrouped away from the infant to consider our options. Perhaps we could remove Samson from the scene by farming him out to Robin's in-laws for a few days, to separate the two dogs and break the predatory cycle. Also, since I confirmed that Samson was showing a predatory response, I would prescribe a course of medication, either the antidepressant amitriptyline or buspirone, which I thought might have some antipredatory action via their effect on serotonin.

Robin and Barry proceeded to follow my advice. They took Samson to their in-laws, retrained and exercised them as I suggested, and began a regimen of treatment with buspirone. The combination was a success on trial visitations. Without consulting me, they brought Samson back into their home after a couple weeks, and he remained calm even when the baby wailed—formerly a sure trigger for his anxious behavior. Robin called me, gushing about the miraculous change.

But when the medication was discontinued after one month, Samson regressed to his old infant-threatening ways. I sternly ordered full-scale safety measures and another course of buspirone. Once again, Samson's behavior improved. This time, after another two months of medication, he was weaned off the drug with no resumption of his old predatory habits. My long-term follow-up indicated that Samson and

Delilah both had finally accepted the new creature as a member of their pack, to be protected and cherished rather than considered potential prey.

Predatory aggression is another fully functional, absolutely normal, and reasonable type of aggression. Reasonable, unless you are the prey, that is. Dogs, like cats and humans, are a predatory species and behaviors for hunting and killing are hardwired in a brain region called the lateral hypothalamus, which regulates appetite and leads to the procurement and consumption of food. Unlike other forms of aggression, the predatory kind does not involve malice or anger. It's just business as usual for a predator on the go. Nothing personal, says the lion to the antelope. The appetitive phase precedes the consummatory phase as surely as spring comes before summer.

House cats perform a mini-me example of a lion's attack. Their predatory aggression is sometimes referred to as the "quiet biting attack." They stalk, crouch, chatter their teeth, and finally run and spring on unsuspecting prey. Their intent is to kill and then devour the prey. But the killing is not done malevolently or in defense. It doesn't involve anger or fear. It's not aimed at imposing their will. It is simply the functional means to a necessary end, which is staying alive.

Modern domestic dogs don't *need* to hunt and kill prey, but they often do so anyway. Some dog breeds with strong prey drive—most terriers, for example—are still prized for their ability to find and kill rodents.

My sister Angela had a German shepherd who liked to chase squirrels. We wondered what the dog would do if she ever caught one. The answer came one day when a squirrel she was chasing got caught up in a tennis net in her backyard. Dog and squirrel came face to face in a standoff. Neither moved for a few seconds, but then the squirrel leaped at the dog's face and bit her in the nose. Big mistake. The squirrel immediately became an ex-squirrel!

Herding breeds also have high prey drive, but the difference is that it's mostly restricted to the appetitive phase, because the consummatory phase of predation was bred out of them years ago by their handlers. Sheepherding dogs who killed and ate their charges were culled and dispatched. That artificial selection has, over time, produced sheep dogs who will do no more than merely nip at sheep.

High prey drive in animals sometimes "dyslexes"—that is, it fixates erroneously on nonprey subjects. Most terriers and sheep dogs, for example, are avid ball chasers. Sometimes, however, that high prey drive is misdirected in unacceptable ways. Dogs may chase running children, joggers, skateboarders, and cyclists. In a few extreme cases, as with Robin and Barry's springer spaniels, dogs view an unfamiliar crying infant as wounded prey, and move in as if for the kill. It is vital that new parents introduce a baby to the family dog under close supervision.

Movement of any kind is a strong trigger for predatory behavior. Dogs often go after passing cars and trucks, which is dangerous for both drivers and animals. I knew of a farmer who crafted a dangerous homemade solution to the problem. He tied a bandanna to the wheel hub of his own car, then drove it repeatedly past the vehicle-chasing pooch (at moderate speed only). The dog would pursue the car, seize the handkerchief in its teeth, and get rolled head over tail. The farmer's tough-love approach broke the animal of car chasing forever, but it could as easily have killed his dog.

Vets prefer a kinder, gentler, more instructive approach. In that regard, a head halter works well. We have owners of car-chasing dogs stand by the roadside with their pet on a leash, head halter in place, and wait for a car or truck to roll by. As the dog starts its shenanigans, we have the owner say, "No" or "Leave it!" as they apply gentle but steady upward tension to the leash until the dog quits. Pressure on the dog's muzzle is thought to replicate a mother dog's correction of her young, so it is biologically appropriate. It works, though the tactic may have to be repeated a few times on different days until the dog realizes that the owner's word is law.

Pet dogs have no need to hunt and kill, because they're fed by their owners. But because prey drive is so deeply embedded, dogs still may express predatory outbursts. It is truly instinctive. Dogs with high prey drive who are not properly exercised will be more likely to engage in annoying predatory behaviors. Those not given appropriate outlets for their instinctual drives sometimes resort to dangerous attacks. Terriers, who were bred to hunt and kill vermin, can be entertained by earth trials, in which they traverse man-made underground tunnels. Their quarry is often a varmint-scented cloth, though sometimes it's a live rat or rabbit in a protective underground cage. In organized barn hunts, dogs search among bales of hay for live rats safely housed in indestructible little cylinders. The rules for humane treatment and protection of the rats used in such trials are (quite rightly) longer than the rules of the game. Shepherd-type dogs can be enrolled in fly-ball classes, or entertained for hours at home with a Chuckit! and a tennis ball. Sight hounds revel in lure coursing. Sporting breeds and scent hounds enjoy tracking.

These activities are not merely occupationally enriching games. They also help fulfill natural hunting instincts and create a happier, healthier, more biologically satiated dog. Such dogs are better and safer house pets.

Cats also enjoy the opportunity to expend their energies, too, focusing on such safe prey as feathered wands or laser mice. Those who don't may resort to what is called a vacuum behavior, in which they act out their instincts by chasing and pouncing on imaginary prey. They are, so to speak, all dressed up with no place to go: equipped with a predatory drive, but with no real quarry in sight. There are numerous indoor toys for cats to exercise on and interactive games for hunting balls, stuffed mice, and other objects.

Some forms of aggression seem to defy classification and treatment. Some may have a genetic origin. Others may simply be normal behavior with the brakes off. We can learn a lot from atypical cases, which can teach us about why and how aggression arises. Investigating the extremes, we can test the limits of what is normal.

One case we had recently at Tufts involved a male Malinois, a short-haired version of a Belgian shepherd that in appearance resembles a German shepherd. The Malinois displayed violent aggression toward his male owner, and before we got involved, bit off half his owner's forefinger. The affected pooch did not respond to any of the usual treatments, strategies that included avoidance, making the dog "work" for all food and treats, and various medications.

Because the dog was unresponsive, we had to dig deeper. Searching the literature, we became aware of a particular genetic glitch in this breed, recently discovered at the UC Davis veterinary school. The culprit seemed to be a faulty dopamine transporter gene, which suggested that trazodone, an old-fashioned antidepressant that can reverse the glitch, might work. We tried it and some desensitization exercises— bingo! Aggression no more.

We had learned from this single, particularly instructive case that behavior modification alone may not always be sufficient to extinguish aggression. Because I'd been trained as an anesthesiologist, in my early career as a behaviorist I naturally turned to medication as a solution to intractable behavior problems involving aggression. At the time, many of the available pharmacological options produced unacceptable side effects such as hyperactivity, or were sedating to the point that the animal was asleep much of its life. Valium, for example, can be addictive and also often has paradoxical results, meaning that it could sometimes trigger the precise behaviors we were trying to prevent, including causing excitement or even increased aggression. I did not want to turn to popular progesterone-like synthetic hormones, either, because they cause unacceptable side effects, such as increased thirst and appetite, hair loss, depression, lethargy, and even cancer.

So I had to look elsewhere for new solutions to decreasing and managing aggression. Something was fueling the fire of aggression, and I wanted to find out both what was kindling it as well as how to douse the flames. To this end, I made diagrams of the pathways and interactions between various neurotransmitters, brain chemicals that convey messages, searching for the points at which a medication might achieve an antiaggressive effect.

All roads led to serotonin, the brain chemical with mood-stabilizing effects. Serotonin is a "neuromodulator." Serotonin causes a spectrum of effects in the brain but in terms of aggression, the more serotonin is present, the less aggression is expressed. Very low serotonin levels are associated with depression and with high levels of aggression. This includes the ultimate form of self-directed aggression in humans— suicide. Robust levels of serotonin are associated with confidence, enhanced social behavior, and reduced aggression. Excess levels of serotonin cause anxiety, as is sometimes seen when a patient is exposed to too high a dose of a serotonin-enhancing drug too quickly.

Amid this welter of indicators, I needed to determine a "Goldilocks" level of serotonin-enhancing drugs to resolve issues of aggression—not too hot, not too cold, just right. All I had to do was figure out which drugs I could safely give to dogs, cats, and other animals, medications that would augment or tweak brain serotonin systems without "going over the top" to achieve my goal of reducing aggression.

My early attempts at treating aggression medically were not as successful as I would have liked, because at the time, the only drugs that were readily available were tricyclic antidepressants. So called because of their three-ringed structure, tricyclics were some of the earliest medications developed to combat depression. These drugs increase serotonin levels between nerve cells by preventing upstream cells from reabsorbing it, thus keeping it at a stable level. Theoretically, this stabilizes mood.

Unfortunately, tricyclic antidepressants also tend to elevate levels of norepinephrine, an excitatory brain chemical, which runs counter to reducing aggression. Increasing norepinephrine may be beneficial when treating pure depression, but higher levels may also lead to aggressive outbursts.

A few dogs that I treated with Elavil, a brand name for an early tricyclic officially called amitriptyline, actually became more aggressive, so I of course discontinued that drug for treating aggression and focused instead on a more serotonin-specific tricyclic antidepressant called Anafranil, a brand name for clomipramine. Despite a few drawbacks, this drug was the stand-out option. Clomipramine enhanced serotonin levels but had much less of an effect on other neurochemicals. In cases where behavior modification alone wasn't helping, I started to include clomipramine in the treatment regimen.

One aggressive dog I saw was a young Rottweiler who was aggressive to his owner. When clomipramine was added into the mix of behavior modification therapy, the aggression practically disappeared. The owner was overjoyed to have his dog behaving normally and affectionately again. Case solved, right?

Not quite. The owner called a few weeks later to report that his dog had demonstrated odd behavior when on a walk and wondered if it could be a side effect. The Rottie had frozen on the sidewalk outside his house, trembling and salivating. I thought that he may have been having a partial seizure, which is certainly possible with clomipramine, though uncommon. Most side effects, including seizures, are dose-related so that was something to consider. This dog was on a regimen 50 percent above the normal dose because the situation was so severe and life-threatening in more ways than one. It may have had the seizure coming on anyway, with or without medication, but to be cautious, I gave the owner two options: stop giving the medication or reduce the dose. Because the treatment was so successful, there was no way this owner was going to give up on it, so he opted to reduce the dose. Fortunately, the dog did not have another freezing attack on the lower dose, and the aggressive behavior did not return. We'd found the Goldilocks dose level.

Soon after successfully treating the Rottweiler and other aggressive dogs, I began to try another serotonin system modifying drug, buspirone, to treat aggressive dogs and cats. Single-dose buspirone had already been proven as an acute antiaggressive measure in vervet monkeys on the Caribbean island of St. Kitts. These monkeys were notoriously aggressive and to study how a drug might reduce their reactivity, scientists first established how they responded to being (gently) prodded with a pole. They hated it and reacted aggressively. Well, who wouldn't? But when the monkeys were given buspirone three hours before the prodding, they took the provocation in stride, barely reacting.

Neither clomipramine nor buspirone produced long-term sedation, a common negative side effect with other psychotropic medications. Both are considered "smart drugs," substances that produce the desired effect with minimal collateral damage. Other purer serotonin-enhancing drugs such as Prozac—the commercial name for fluoxetine—became widely available and more affordable a few years after my early cases of aggressive dogs, so I included them in treatments. Fluoxetine worked well to reduce aggression in dogs, cats, horses, and even a parrot or two. Serotonin-selective drugs like fluoxetine were clearly superior to clomipramine, and I could use them along with buspirone if necessary. Over time, fluoxetine became more and more the treatment of choice across the veterinarian community. Our pets officially joined the so-called Prozac Nation.

In a recent case, I used a one-two combination of Prozac and Buspar, not so much for fear-based aggression as abject terror. A young adult German short-haired pointer, Lita, refused to leave the Boston apartment of her owners. The trigger for her behavior was unknown. She might have been traumatized by some scary noise while she was out walking. In any event, fluoxetine alone did not cut it, so I added buspirone. The result was a spectacular success. Lita pranced down the corridor to the elevator, as happy as she had ever been to go out. No more balking, no more struggles to get her out for a walk. Just a happy dog and happy owners. The final outcome was even better. After about three or four months, I was able to discontinue the medications and the improvement to Lita's behavior remained. I call this effect "pharmacologic desensitization." If subjects experience a situation without fear many times, they are able to learn that it's safe and nonthreatening.

I continued to root around in canine brain circuitry, looking for ways to prevent aggressive outbursts. Because neurotransmitters are instrumental in triggering the fight-or-flight response in animals, I searched for ways to block the action of pro-aggressive brain chemicals. If I could impede even some of their effects, I figured I could reduce the agitation and mental foment that often precedes aggressive behavior. I thought that perhaps a beta-blocker called propranolol might work, since it blocks some of the effects of a neurotransmitter called norepinephrine, but it turned out to help my animal patients only slightly and only in some cases.

Despite my standing concerns about using drugs in the Valium family because of their side effects, I cautiously started to employ them to treat aggression. I used Valium itself, which is a brand name for diazepam, as well as related medications, alprazolam and clonazepam, which are marketed as Xanax and Klonopin respectively. Through trial and error, and working out how a particular animal might react to a drug, I got some good results in reducing aggression, especially in aggressor cats.

A pair of constantly feuding cats came to see me at Tufts. The conventional first level of treatment is to separate and gradually reintroduce them to each other—called systemic desensitization. This can click the reset button for the relationship, but since they'd become entrenched in their warfare, I also tried various medications to facilitate peaceful interactions between the pair. Nothing worked. These feline fighters were displaying territorial aggression: they had dug in and were not open to detente.

It might take two cats to tango, but one of the pair clearly danced the part of the lead aggressor. After almost a year of trial and error, I tried Xanax to reduce the bully cat's anxiety and aggression. It worked! Since then Xanax has come to the rescue in placating other angry cats, especially those who do not care for another housemate.

In even more persistent cases of aggression, I resorted to using heavy-duty antipsychotic medications, including Haldol, the brand name for a dopamine blocker called haloperidol. Simply put, dopamine acts as the chemical connection between thought and action. Without sufficient levels of dopamine in our systems, we slow down and have movement disorders, as in Parkinson's disease. With no dopamine in our systems at all, we become immobile, the sad condition of the patients described in one of Oliver Sacks's books, *Awakenings*.

When we block dopamine with drugs such as haloperidol, animals become less aggressive, but the side effects are unacceptable, and include a reduction in activity level and strange movement disorders called dyskinesias. Human patients who take strong antipsychotic drugs for schizophrenia or manic depression for a long period can develop these involuntary grimaces and jerky large-muscle movements, or dyskinesias, as can dogs and horses, although animals generally do not show bizarre facial contortions as people do because they lack the same facial muscle structure and innervation.

I treated an aggressive Doberman with a dopamine blocker and warned the owner that she might see a dyskinesia in the form of head bobbing. I guess I wasn't clear enough that this side effect is a bad thing and she should stop the medication if it happened. A couple weeks later she called to say that her dog's behavior problem was much improved, but that the animal was now head bobbing.

"Why haven't you discontinued the drug?" I asked.

"But you told me that's what might happen!" she replied. Of course, I had her halt the medication immediately. The head bobbing stopped, fortunately, and the aggressive behavior returned, unfortunately, but we moved on to treat it with the logical next choice, a serotonin-enhancing drug, which worked for this Dobie.

I wasn't the only one working on the pharmacological control of aggression in clinical patients. The psychiatrist Dr. John Ratey treated, among others, violent human offenders in Medfield State Hospital, Massachusetts. In a talk at Tufts Department of Psychology, Dr. Ratey discussed the types of medications he used, and I was gratified to hear they were almost identical to the ones I had come up with on my own. Dr. Ratey used antidepressants, like clomipramine, to treat aggression in people, as well as desipramine, an older tricyclic drug. In a subsequent appearance on the TV news show, *20/20*, one of Dr. Ratey's desipramine patients said he wished he had been treated earlier, before his wife divorced him, before his children stopped speaking to him, and before people at work dreaded seeing him entering the building.

Dr. Ratey also spoke highly of buspirone, and Valium-type drugs. The beta-blockers he used to treat his aggressive human patients seemed to help by reducing tension in muscles, something like a pharmacological back massage, rather than acting primarily on the brain. Dr. Ratey found this out by giving his patients a beta-blocking drug that does not cross into the brain. The medication still worked.

After the talk, Dr. Ratey and I chatted about the similarities between responses to medication in my veterinary patients and his human ones. We found that we spoke the same language; we agonized together about the lack of any formal antiaggressive treatment for man or animals.

There are almost a million aggravated assaults in the United States each year, a number that does not include instances of violent rape and murder, but aggression is not even formally recognized as a condition in the Diagnostic Manual of Psychiatry. There is no medication formally recognized or FDA-approved to treat aggression. That's because drug companies do their best to avoid what they see as "unnecessary" risks of lawsuits.

One example of an aggressive dog who did very well with behavior modification paired with pharmacologic treatment was Lucky, a castrated male Walker hound mix. When he was two years old, he had been found wandering the streets, severely emaciated and with several broken ribs. Adopted out of a shelter, he quickly exhibited fear aggression toward people and other dogs.

Lucky would lunge, bark at, and try to bite anyone who approached him, including his owner's mother and brother who lived in the same house. He had also attacked two other dogs who lived in the home, a cocker spaniel and a Rottweiler. Lucky also chased passing cars, a situation that got so bad the new owner couldn't walk him on the street.

As I worked with Lucky, I taught the owner the importance of avoiding triggers. We switched the dog to a Gentle Leader type of head halter, which applies pressure to the back of the head and under the throat, not the neck. In general, animals are more easily controlled by the head than by the neck. I also instigated a behavior-modification program that was basically opposite to the one that had worked for my dog Rusty, to whom I'd given free food and treats to boost his confidence. For Lucky, instead, the motto became "Nothing in life is free." Lucky had to work for every piece of food, for every pat on the head, and even for his freedom. I advised his owner to get Lucky a close-fitting antianxiety wrap, to increase his exercise, and to set clear boundaries of acceptable behavior.

Things improved a bit, but not enough, and his owner considered euthanasia. Lucky had bitten the owner's brother and they all worried he'd harm someone else. I couldn't let that happen.

Lucky's life was on the line. I started him on Prozac as a mood stabilizer and also prescribed an "as needed" dose of clonidine, a new treatment for aggression that I had borrowed from human medicine, which works by reducing the release of norepinephrine (the excitatory hormone) in the brain. This could temper Lucky's fight-or-flight response in particularly challenging circumstances.

The results were encouraging and, six months later, Lucky's owners reported a 50 percent reduction in his aggressive behavior. I wanted even more improvement, so we upped the dose of Prozac. This finally did the trick. One month after the increase, Lucky was able to go to doggie day care, he no longer attacked the family or the family Rottweiler and spaniel, and even played with them occasionally. The formerly fearful and aggressive dog was now very trusting of his owners, and he even allowed strangers to approach him.

It took a period of trial and error to discover the proper treatment for Lucky, as it does with many dogs—and for people. There is no "one size fits all" when it comes to behavioral pharmacology. My motto is, "Nothing works all the time," a guiding principle that has stood the test of time. It's humbling to acknowledge that the brain is a very complicated place and no one knows all of its mazes and interconnections enough to predict exactly what adjustments are necessary in aggression cases—or any other behavior case for that matter.

Training was crucial to Lucky's rehabilitation. The combination of training and medication saved his life. Prozac has saved the lives of millions of people, and I was glad that we had jumped the species barrier with it and had saved Lucky's life—and would go on to save others.

CHAPTER TEN

Animals Who Fear Too Much

Anxiety and Panic Disorders

If having a soul means being able to feel love and
loyalty and gratitude, then animals are better off than a
lot of humans.

—JAMES HERRIOT

When Mindy, an eight-year-old spayed female dog, came to me at Tufts, her owner wrote down "stranger anxiety" on the admission survey as a description of her dog's problem.

Mindy's past made her ripe for behavior issues. She was adopted from a shelter when she was between three and five months old. She had then undergone a two-month quarantine period because of parvovirus infection and had probably not been socialized properly when young. Until the time of her adoption, Mindy had negative or at least inconsistent interactions with people and her isolation in quarantine at a formative time of her life had exacerbated her fearfulness and insecurity.

All these influences came together to produce a dog who was deathly afraid of all strangers. It was pitiful to see Mindy in the consulting room. To start with, she stayed close to her kindly owner or pressed herself close against a wall. Her whole demeanor was one of mistrust. She eyed everyone in the room with furtive sideways glances. When her owner moved away from her, she hid under my desk and refused to come out, not even when lured with offers of food.

Because of her overwhelming fear of strangers, Mindy had been treated with mood-stabilizing medication for most of her life. Her owner tried to wean her off the medication at one point, but Mindy had gotten much worse and had to be put back on it again. My job was to optimize Mindy's medications, to tweak her drug regimen, to try to make her a little more social. I also advised her owner on how to modify his pet's behavior and coax Mindy out of her shell. Mindy somehow had to learn that not all people are to be feared. During the office visit, we had some success with Mindy eventually taking food treats from her safe place under my desk.

Although she improved considerably after our first encounter, Mindy still remained far from confident in the company of people. I should have predicted such an outcome, because, sadly, fears once acquired are not easily forgotten.

Anxiety can express itself in many different ways, including situational anxiety, social anxiety, stranger-induced anxiety, and generalized anxiety. There are as many brands of anxiety as there are of fear. Anxiety also underlies obsessive-compulsive disorder. For such a varied condition, it's odd that the diagnosis of anxiety can still be controversial.

I once read a behavioral pharmacology textbook that confidently labeled fear as a major component of anxiety. That's obvious, right? Yet many experts take exception to this as an overly liberal linking of the two concepts. But I think it has some merit. Fear and anxiety are related in their underlying neural mechanisms, though we—and animals—obviously express them somewhat differently.

Fear and anxiety can be situational and temporary, or ongoing and debilitating. Fear occurs when a potentially noxious stimulus is actually present, right there in front of us—animal or person. Anxiety, on the other hand, is an anticipatory type of fear. A pet fears that something bad may happen, or that something highly desired, like an owner's return, may not come to pass. With anxiety, a direct stimulus is not present, so we call it *objectless*.

If a primate sees something scary, he naturally reacts with a "fear response." It's a natural protective reaction designed to prepare the animal for fight or flight. The stimulus might be a snake, fear of which is innate and anchored deep in our biology. As a primate, you really don't want to step on that possibly poisonous snake in the grass or have it drop on you from a tree.

Primates and many other animals are genetically predisposed to react with fear to various triggers, including snakes in the grass, hawks circling overhead, and spiders. Nature knows best. So we have genetically determined causes of fear, and we also learn certain fears. When an animal is scared by an event or object, memories are stored for future reference. This "fear learning" is adaptive, in the sense that it is good for an individual's survival. All of us mammals share this ability to learn from experience—and the tendency to react fearfully to and want to avoid a remembered negative experience.

Fear responses include the proverbial fight-or-flight reaction, freezing in position so as not to attract attention (as does a deer in headlights), hiding from the source of terror, or seeking the protection of another person, animal, or group. Another fear response goes by the tongue-twisting label of "thigmotaxic behavior." With thigmotaxic behavior, a common strategy for rodents, animals hug the borders of a containment area, putting their backs to the wall, so to speak. Unfortunately, not all strategies work well for animals in the modern world. Freezing in the headlights of a car, for example, does not bode well for either the animal or vehicle.

In addition to these outward or operant behaviors, as these fear responses are called, automatic internal responses occur to prepare the animal to deal with the threat. These include increased heart rate, blood pressure, respiratory rate, hair standing on end, increased muscle tone, sweating, and, in dogs, salivation.

These fearful operant and autonomic responses are similar to what happens when an animal is anxious, but less well defined. Classic manifestations of anxiety include attempting to escape from an acute anxiety-promoting situation, avoiding contact with a disquieting person or other animal, hypervigilance, nervous pacing, stomach upsets, and digestive disorders.

Both fear and anxiety can be functional, that is, they can help an animal or person be properly prepared for challenges that they encounter. But fear and anxiety can also become overreactions. Excessive and dysfunctional fearfulness, phobias, or excessive anxiety caused by overactivity in certain brain circuits, can interfere with daily life and cause debilitating psychological problems.

Many people ask whether animals can experience anxiety, to which my answer is a resounding yes. Of course, you can't ask an animal whether it is feeling anxious and hope to get a definitive answer. But by all other assessments—by outward behavior and measurement of internal responses—anxiety in pets is as real an entity as pain. Anyone who lives with animals knows this. Pet owners more readily accept that animals can feel anxiety than some purist behavioral scientists.

Because of the constraints of Morgan's Canon, some scientists balk at the concept of an animal feeling anxiety. *Real* scientists, they believe, should always find a way of explaining behavior in basic mechanistic or reflexive terms, rather than attributing what they see to a cognitive process.

Research data on brain function, unavailable to Morgan when he formulated his canon, now show that the neurological centers governing fear and anxiety are similar in people and animals. The amygdala, the brain's Grand Central Terminal for both fear and anxiety, lights up on PET imaging scans in anxious and fearful animals and people.

A long-term memory center, the hippocampus, is also involved in propagating the response. Connections between the amygdala, hippocampus, and hypothalamus facilitate release of stress hormones, like epinephrine and cortisol. Epinephrine increases heart rate, blood pressure, and the caliber of the respiratory passages. Pupils dilate widely. Depending on the species, the mouth of some animals goes dry because of lack of saliva, and dogs drool a thick viscid saliva. In addition, endorphins are released in reaction to fear.

So in both people and nonhuman animals, the same neurotransmitters are released in brain regions that deal with anxiety and similar behaviors result. The similarities of people's and other animals' responses are quite clear to me. Morgan's Canon be damned.

Why would nonhuman animals not be self-aware? Why would they be incapable of projecting bad outcomes for themselves? Recall the Harvard research studies that strongly suggest that dogs possess "theory of mind," demonstrating that some other creature knows something that they don't know. Other studies show that dogs can be jealous and that dogs laugh, making a happy huffing sound when they are amused. Kenneled dogs calm down and stop barking when exposed to sounds of other dogs huffing with laughter.

Finally, dogs have been reported by scientists at Bristol University to have either optimistic or pessimistic character traits. Joy is the primary emotion underlying optimism and sadness is the primary emotion underlying pessimism. By this score, laughter should occur more often in optimistic canines and anxiety may occur more frequently in sadder, more pessimistic dogs. Which is exactly what the research shows.

Put a dog in an anxiety-promoting situation, or leave him alone in a strange place, and observe him pacing, whining, and breathing hard.

Measure his heart rate, blood pressure, and stress hormones such as cortisol and endorphins, and you will find them all elevated. Then reverse this whole situation with a known antianxiety medication.

Other than anxiety, what explanation can there possibly be for the similar brain activity and physical behavior? To me it's a no-brainer, so to speak. Dogs can be anxious, cats can be anxious, horses can be anxious. All animals can be anxious.

In people, a common type of anxiety occurs as a panic attack, in which intense feelings of fear or discomfort are accompanied by several physical signs of distress, like sweating, trembling, and shaking. These attacks can be so severe that sufferers often believe that they are going crazy or are about to die. They feel an overwhelming instinct to flee from wherever they are, but sometimes are so overwhelmed that they pass out. Panic attacks can arise unexpectedly or they can be provoked by certain environmental factors.

I am not sure that dogs have panic attacks that occur with no specific trigger, but they surely experience such attacks in situations that scare them. One dog I treated had "panic attacks"—as described by his owner—several times a day.

Spock earned his name because of his large, batlike ears. Unlike his namesake, Spock was emotionally over the top, and illogically so. He would become totally hysterical several times a day, pacing, panting, and vocalizing for several minutes at a time. In this specific case, the trigger turned out to be the remote sound of any form of power equipment in a nearby neighbor's house.

Another dog exhibited the same type of behavior several times a day, but only in his owner's home, never when at a friend's house. The trigger in this case, determined later by a process of exclusion, was the owner's own parrot, whose screeching terrified the poor dog.

Another pooch reacted to the sound of steam blasts emitted from a clothes iron.

Some people who have panic attacks also have agoraphobia. They may panic when home alone, as the definition includes being in any inescapable situation, not simply being in a wide open space. The separation anxiety that dogs suffer resembles a panic attack, with affected dogs showing physical signs of severe anxiety.

Separation anxiety occurs in some 15 to 17 million of the nation's 78 million dogs. Certainly some dogs seem to inherit a tendency to develop separation anxiety, so nature possibly plays a part, but inadequate nurture at an early stage of life seems to be the most important factor. At Tufts, when a dog comes in with separation anxiety, we ask the owner a series of simple questions regarding the pet's early upbringing.

- Was the dog separated from his mother and littermates too early, i.e., before eight weeks?
- Has he had multiple owners?
- Did you acquire him from a shelter?

Almost always, positive answers to these questions indicate that the dog had a dysfunctional early puppyhood. Young dogs should not have to face these stresses. There's a basic rule: The more consistent the attention paid to a youngster in an early sensitive period of development, the more confident the adult animal will be. The opposite also tends to be true. Less attention and more disruptive environments lead to anxiety-related behavior issues. It's the same for people.

At least half the dogs with separation anxiety come from a shelter. When this is not the case, I can usually identify periods when the dog, as a puppy, was isolated, frightened, or psychologically traumatized.

Two responses to psychological trauma are common in people: they either seek to take greater control of a situation or they lose self-esteem or confidence. Dogs with separation anxiety have little confidence. They can make for the sweetest pets, following at their owner's heels everywhere around the house. They are generally worried or anxious, offer frequent appeasement gestures, and live in mortal dread of their owner's absence. Whenever their owner does prepare to leave, they become extremely anxious, pacing, whining, hiding, or shaking with fear. After the owner leaves, they pace, whine, bark, or howl, and sometimes panic. Those who try to escape may damage doors, windowsills, or blinds. Others act out their frustration with displacement behavior, destroying couches and cushions, cupboard doors, or other wholly innocent targets. Some dogs throw up because of the anxiety. Few pay any attention to food, displaying so-called psychogenic anorexia. Many have "accidents," soiling the house.

Unwitting owners quite often attempt to resolve the hallmark destructive behavior by crating their dog. That may solve the owner's immediate problem, but it does not usually address the pet's underlying issues. Some dogs become even more hysterical when crated, damaging themselves and the crate in sometimes successful attempts to escape their confinement. In the midst of crate panic, it's as if the pet entered the phone booth as Clark Kent and emerged as Superdog, developing the strength of ten. Superdog will bend and buckle the metal bars of crates as if they were made of putty. In the process, they may chip teeth, bloody their mouths, or break off claws. Owners who try the confinement solution might return to find their dog still in the crate but bathed in blood and standing in a pool of saliva. One crated dog barked so hard and for so long that he jumped his crate across the floor with the vibrations. Reaching the door where his owner's purse was hanging, he sucked it in through the bars and ripped it open, strewing its contents—makeup included—around the inside of the crate. Then he continued to bark and managed to jump the crate back to its original position. When the owner came back, she was stunned with what she saw—the torn purse inside the crate, not where she had left it, and the dog enthusiastically welcoming her home with lipstick all over his face. It was a sight for sore eyes.

Separation anxiety can be a burden for owners and neighbors, who may complain about the dog's continuous barking when alone. Owners often have costly property damage. Many owners are angered by the behavior and damage, but others feel sympathy for their pet. As bad as all this is for owners, it is worse for the dog. Many of these traumatized pets are exiled to shelters, some to face euthanasia.

One dog with the most severe separation anxiety I have ever seen leaped clean out of a window—except the window was shut. He broke right through the glass and fell to the ground below. The neighbors heard the crash and ran outside to see what was going on. When they saw the broken window smeared with blood, they concluded a vicious crime had been committed and called the police. The owners later found the dog wandering, cut up and bleeding profusely. When I saw the dog, he was covered with bandages like an Egyptian mummy.

The causes and symptoms of feline separation anxiety are similar to those of dogs. Cats vocalize and pant and pace. They don't usually make enough noise to disturb the neighbors and are not powerful enough to trash micro blinds or destroy doors and window moldings. Anxious urine marking occurring *only* in an owner's absence is a cardinal feature of the condition in cats.

Children get separation anxiety, too, a state of affairs that would be easy to predict even if it had been never before described. Separation anxiety is one of those situational anxieties that are well-recognized in human kids, especially when it comes time to leave Mom and go to school for the first time. My now fully grown and wonderful daughter Keisha had a bout of separation anxiety when she was very young, when we enrolled her in day care. Her symptoms of crying, inappetence, and panic were similar to those I see in my canine patients. She was too young to try to escape day care in order to get to her mom but otherwise the parallels were undeniable.

To prevent separation anxiety for children and dogs, it is best to train independence gradually, in order to avoid the sudden loss of an attachment figure. My daughter recovered from her separation anxiety and is now confidently living away from home, having graduated from medical school. Puppies are all too often taken too early from their mothers in puppy mills and stuffed into pet stores. We can help dogs with separation anxiety by avoiding prolonged absences from home, independence training, environmental enrichment, and, when necessary, antianxiety medication.

Another kind of anxiety is social anxiety, which the chronically cowering Mindy displayed. Genetics probably plays a role in setting up animals for this condition, but improper socialization leading to lack of confidence in social settings is the precipitating factor.

One study found that cats not exposed to people in the first seven weeks of life were always on edge in the presence of people. This sensitive period is crucially important to normal feline development, too. Certainly cats who have not been socialized can form close relationships with one or two trusted individuals, though total acceptance of people apart from those of their inner circle is not a realistic expectation. When these cats are around even well-intentioned, nonthreatening strangers, their anxiety is practically palpable. Mostly they simply hide under furniture and remain stressed and hypervigilant until the person they see as a potential threat is long gone.

Socially anxious canines also become stressed in the presence of unfamiliar people. They are not aggressive, just anxious. Dogs may squat and urinate around strangers or even their owners. Although commonly misunderstood by owners, submissive urination is a gesture of appeasement, designed to assuage the perceived threat. Getting angry or impatient with a dog who does this makes the problem worse.

Dogs with social anxiety dread crowds, even when nothing bad is actually happening. Underexposure to people in the sensitive period of development causes dogs to develop social anxiety. A severely circumscribed life, as occurs in brood bitches in puppy mills, will totally destroy self-confidence in dogs and lead to social anxiety.

Adult humans may suffer from social anxiety and there is some evidence that genetic factors play a role in its development. But, again like dogs and cats, environmental factors are more intimately involved. Children with social anxiety dread all interactions in a public setting, often worrying that they will be judged negatively by other people. They also are afraid of strangers. Stranger anxiety typically occurs in children under one year of age when they are looked at or approached by a total stranger. It is so common for children to have this fear at this stage of life that it is considered almost normal. At this early point in development, stranger anxiety is certainly not considered a phobia.

Young puppies and kittens may go through a similar stage. A stranger in the house might cause fear, flight, hiding, or overly extreme and abjectly submissive reactions. With good management, this is a fleeting phase of development. In both people and pets, if stranger anxiety persists into adulthood, it is termed social anxiety.

Generalized anxiety disorder (GAD) is a psychiatric diagnosis that we frequently use to describe some generally nervous or fearful pets. The fourth edition of the DSM states that, for diagnosis of this condition, people must experience chronic anxiety and excessive worry, almost daily, for at least six months. They also must have three or more defined symptoms, including "restlessness or feeling keyed up or on edge, being easily fatigued, difficulty concentrating or mind going blank, irritability, muscle tension, sleep disturbance."

There is no doubt that GAD also exists in pets. Some dogs seem locked into a perpetual state of worry and display ongoing anxiety in three areas: anxiety around people, anxiety concerning certain sights and sounds, and one or more situational anxieties. This is a behavioral "full house" as far as anxiety is concerned. These animals may develop fears to the level of phobias or manifest the anxiety-based condition of obsessive-compulsive disorder.

Brood mothers from puppy mills frequently have the behavioral full house of anxiety that I classify as GAD. Practically afraid of their own shadows, brood bitches from a puppy mill will cower around people, seemingly terrified of everyone and everything. They often remain silent, never barking at all, and do not wag their tails. They don't know how to play. Most do not recognize or climb stairs, have no idea about cars or car travel. They are, in general, totally wretched. Their cruel fates have robbed them of their normal canine selves.

Fortunately, something can be done to address such ongoing anxiety. It takes time, patience, and positive training methods. In the right hands, GAD dogs can be usefully rehabilitated in about a year. They are not "cured" in the absolute sense of the term, but they have been coaxed out of their fear and into something resembling normal doghood.

About twenty years ago, I was talking to my next-door neighbor, John, in his front yard. John told me he had been experiencing anxiety as a result of a number of unfortunate experiences occurring simultaneously. He had been trying to hold down three jobs and become unable to sleep properly as anxiety spilled over into the nighttime hours. John found himself caught in a catch-22. Because he couldn't sleep, he couldn't do his work well and feared losing his primary employment as a school chemistry teacher. This made him worry even more. His doctor put him on the antianxiety medication buspirone, which was brand-new at the time and it had worked beautifully. Now he could sleep and handle his multiple jobs. Because he felt better, he worried less. He was a new man.

Using buspirone in a veterinary situation was somewhat novel at the time, but when I found that it worked as an antianxiety treatment in my animal patients, I decided to apply for a US patent. The patent examiner, however, was apparently a man ahead of his time. He declined the patent application. The grounds? Such a use was "obvious to one skilled in the art." Of course an antianxiety treatment for people would also be effective in animals. The idea was too baldly apparent to warrant protection. The examiner turned out to be an early One Medicine proponent, probably without even being aware of the fact.

There was one part of the application that he did approve. I was granted a patent for the use of buspirone to treat anxiety-related urine marking in cats. Considering the logic the examiner was applying, it was odd that even that claim was allowed.

The anxiety-related conditions that occur in people and the equivalent ones in animals, all respond to similar behavioral modification therapy and identical medications. Buspirone and serotonin reuptake inhibitors, such as Prozac, Paxil, and clomipramine, are effective for treating anxiety in people as well as dogs, cats, and horses. Two of these medications, fluoxetine (human brand name Prozac) and clomipramine, are licensed by the FDA for the treatment of separation anxiety in dogs. It moves me to ask a question of the scientists who adhere to Morgan's Canon: What part of One Medicine don't you understand?

The bottom line regarding animal anxiety is that, if it looks like anxiety, has similar internal changes as the ones that occur in human anxiety, and can be successfully treated with the same medications as people with anxiety, then it probably is anxiety.

Help for treating animals suffering from anxiety came from an unexpected quarter. One of the most celebrated people with autism is the scientist, animal activist, and best-selling author, Temple Grandin.

When I worked with Temple a few years ago, I learned a lot about how to reduce anxiety and stress reactions in animals.

I first met Temple when she was a PhD student at the University of Illinois, researching environmental enrichment in young piglets and its effects on brain development. An expert on farm animal behavior, especially the behavior of cows and horses, Temple had risen to become an academic star and visited Tufts to give a lecture.

Temple's relationship with animals is in some ways otherworldly. She seems to intuitively understand animals and their behavior, and she is able to build up nonverbal rapport with almost every creature she meets. If she sits down in the middle of a field of cows, the cows will be drawn to mill around her. They appear to recognize her empathy, and they react accordingly.

Temple has high-functioning autism—sometimes referred to as Asperger's syndrome. Her remarkable skill in understanding animal behavior has made her, perhaps paradoxically, one of the world's most sought-after designers of slaughterhouse facilities.

Her philosophy is that, since people are always going to eat animals, we should see to their welfare and, when their day comes, ensure they experience as little stress as possible. Not only is it the humane thing to do, but stress can trigger surges of adrenaline and other stress hormones, which can negatively affect the taste of meat. A careful, respectful send-off works for both the stock animals and the meat purveyors.

Away from their home barns or pastures, animals clearly feel great amounts of fear, uncertainty, even terror. Temple has noted the calming effect of controlled restraint of animals with mechanical devices. For example, in slaughterhouses, pigs are transported from point A to point B on a conveyor belt and squeal in fear. When they are restrained in a V-shaped contraption called a V-trough, however, their squealing is effectively eliminated. The effect of the V-trough seems profoundly sedating. Even when the workers leave the plant for lunch and turn off the machinery, any pigs suspended in transit in the V-troughs remained relaxed and silent, almost falling asleep.

At Tufts, we used V-troughs to calm and restrain pigs prior to anesthesia. A raucous, violently struggling 100-pound swine is a tough animal to control, as county fair greased pig contests can attest, but once installed in a V-trough a pig becomes serene, almost mesmerized. The effect is as dramatic as hoodwinking a bird, hypnotizing a chicken, hobbling a horse, or scruffing a cat.

Babies, too, are soothed when swaddled tightly. When she herself was young, Temple did not like being hugged by well-meaning relatives, although she felt comforted when she wrapped herself tightly in the blankets of her bed. This type of gentle, persistent, and controlled pressure generates feelings of well-being.

When in college, Temple designed and developed a squeeze machine for herself, a boxlike structure in which an average person can sit, with inflatable pads that can be controlled to exert pressure on the torso. The effect is calming but not stifling. With just the right amount of pressure for just the right amount of time—about twenty minutes—Temple would emerge from the machine relaxed and happy, able to go through her day with less stress and with a lot more focus. She shared this technology with various autistic centers around the country so that they, too, could use them in therapy for autistic children. Dr. Oliver Sacks, who wrote about Temple Grandin in his book, *An Anthropologist from Mars*, tried Temple's machine and reported feeling perfectly relaxed afterward.

Just what exactly was going on, in terms of biochemistry, when a pressure is calming? After her talk, Temple and I discussed her V-troughs and my work with cribbing horses. The cribber's repetitive actions cause the brain to release endorphins, which help the horse cope with the stress of confinement. Endorphins, natural morphine-like substances released by reward centers in the brain, cause feelings of warmth and pleasure.

Temple asked, "Do you think that the pressure effect of the squeeze machine releases endorphins?"

"Could be," I said.

"Maybe that's why I find the squeeze machine addictive," she said. "I make a point of limiting my use of it to twenty minutes a day. I tell that to people in the autistic centers, too, that they shouldn't let people get too fond of it. It's easy to get addicted to a behavior. That's what happened with your horses, isn't it?"

"Do you think we could run an experiment together?" I asked. "We could try blocking the temporary squeeze-induced bliss in pigs, for example, using an opioid antagonist."

Temple liked this idea. "It's easy to work with pigs," she said. "You've got some on the campus, haven't you?"

"We have pigs in a place called Hog Heaven. Perhaps they would volunteer to be our subjects?"

"They may even enjoy it," Temple said drily. "I'll draw up plans for a pig-size squeeze machine and you can have the machine made. Then we'll run a score of pigs through that machine, give half of them saline for control, and the other half we'll give your blocker drug . . ."

"Naltrexone," I said.

We had a plan. That is, as long as the pigs agreed.

A few days later, Temple's blueprint plans for a pig squeezer arrived, a marvel of draftsmanship. One of Temple's many extraordinary gifts is an ability to draw up and read blueprint designs as if they were three-dimensional objects. Her drawings were works of art.

I gave the plans to the university carpenter, who fabricated a very fine-looking portable squeeze chute for pigs, a padded V-trough with a lever. When the lever was pressed, the sides of the V came closer together to produce gentle pressure on each side of the pig's body. With the machine completed, we were ready to try our experiment. Temple flew in from her home in Illinois.

The temperature was already in the midnineties that summer day. The piggery at Tufts is buried deep in the woods surrounding the campus. I slid open the door to the piggery and we struggled to get the squeeze machine through the small door, eventually managing to position it in the center of the aisle that divided two rows of pens. By this point the pigs were squealing in anticipation. Human visitors usually meant food.

Without any preliminaries, Temple hopped into the first pig's cage, grabbed the animal by its hind legs, and lifted it onto the scale. This was no easy feat. Each piglet weighed sixty or seventy pounds. But Temple displayed impeccable pig-handling skills.

"Pig number 345, sixty-five pounds," she hollered. "Ready for Solution A."

I grabbed a prefilled syringe containing "Solution A," aka saline, and injected it into the piglet's buttock. The needle was so fine the pig didn't seem even to notice the injection. Temple put the pig back in its pen and moved on to the next one.

"Pig number 581, sixty-two pounds," she called out. "Ready for Solution B."

"Solution B" was actually naltrexone, which blocks the action of endorphins in the brain. The experiment proceeded until we'd injected another ten pigs with either Solution A or Solution B.

About an hour after we'd begun, we returned to the first piglet. We attached ECG sticky pads onto the pig so that we could get an instantaneous measure of the animal's heart rate and its rhythm. Temple heaved the animal up and over the bars of the pen, placing it in the padded V-trough. I engaged the lever to apply pressure, and at the same time hit the start button on a stopwatch.

We observed the pig closely for signs of relaxation, like fluttering eyelids, slow breathing, and loosened muscles. Just as Temple had predicted from her slaughterhouse observations, and just as I had noticed when I anesthetized pigs, each animal tested gradually began to relax into a soporific state once the pressure was applied. At the point of maximum relaxation, which came usually between two and five minutes, I hit the stopwatch button again and noted the time.

The first pig, who had been given Solution A, the saline, took 110 seconds to relax. Unfortunately for this blissed-out piglet, we then rudely awakened and hoisted him back into his pen. Clearly none the worse for wear, he rooted around in the straw bedding, grunted a few times, and then lay down, perfectly happy.

Temple grabbed the second piglet and we repeated the whole procedure. This pig had been given the drug that stops the endorphin rush. It remained alert in the V-trough, grunting and tense, for much longer than the first pig.

"I think it's working," Temple said.

When the second pig finally drifted into a dreamy state, I pressed the stop button on the stopwatch at 300 seconds.

"That's nearly three times longer than the first one," Temple said.

On we went until all the pigs had been tested. By the end of the day, our hair smelled of pigs, our hands smelled of pigs, our clothes smelled of pigs. Flies buzzed around us excitedly, unable to decide who to settle on, us or the pigs. But we had very interesting data to study, so we didn't mind.

The next day, we used a different barnload of squealing piglets. We wrangled, injected, pressurized, and recorded, until we once again smelled like two large ungulates, and had additional data to analyze.

Temple flew back to Illinois, and called me the next morning, having entered the study data into her laptop on the flight home. She had also calculated the "means and standard deviations" of the relaxation times, comparing the results of the two groups of Solution A and Solution B pigs.

"Guess what?" Temple said. "Just as we thought. There is a significant difference."

We had clearly proven that endorphins are involved in the calming effect of controlled squeezing. The pigs that received the saline injection relaxed much more quickly than did the pigs given the endorphin-blocking drug. We had taken a significant step forward in understanding how to alleviate stress, both in animals and people.

Autistic children seem to have roller-coaster-ride levels of endorphins throughout the day. When levels are low, our results suggest, the squeeze machine can help boost endorphin release and normalize their feelings. High levels of endorphins though may be responsible for the frantic head banging that is sometimes a feature of low-functioning autism. Endorphin blockers like naloxone are sometimes helpful in reducing this dangerous behavior.

Hugging may generate endorphins, too. It is a sign of affection, and may send a message through the medium of endorphins. Dogs who have generalized anxiety and thunderstorm phobia are now regularly suited up in tight-fitting vests or T-shirts, which produce about a 50 percent reduction in the signs of anxiety during storms. For dogs with separation anxiety when left by their owners in a kennel, vests lessened the rate of increase in heart rate.

While V-troughs, chutes, swaddles, and tightly wrapped arms are not exactly identical, each provides similar psychological benefits and comparable effects. Controlled pressure, both in animals and people, releases endorphins that lead to feelings of inner well-being or peace. The molecules of our emotions stress or comfort us in virtually the same ways.

CHAPTER ELEVEN

Dogs Who Hate Bugs and Storms

The Trouble with Phobias

May all beings be free of suffering.
—BUDDHIST PRAYER OF COMPASSION

Mabel, a delightful one-and-a-half-year-old spayed black Labrador retriever, was a typical happy, carefree dog living a virtual life of Riley in northern New England. But a month before she was brought to my clinic, her owners had taken her swimming with another canine friend. It was June, the weather was perfect, and the dogs had much fun splashing in and out of the water. Cue the rise of ominous music here.

At some point, her owners suddenly noticed that Mabel was not around. They searched and found her a few minutes later, buried beneath blankets in the back compartment of their crew cab truck, hiding as if in fear of some imminent threat. As the owners drove home, they noticed Mabel licking her belly and scratching a lot. At the time, they didn't worry that much about her behavior, concluding that if something was bothering her, Mabel would soon shake it off.

It didn't happen. From that day forward the poor Lab was a changed dog. Mabel was now always on the lookout. "Hypervigilant" was the term her owners used. Her tail was constantly tucked. She circled anxiously, as if anticipating that something awful was about to happen. At the time of the consultation, her owners reported that since the mystery incident, Mabel had not wanted to go for walks.

I had seen this type of problem before and was pretty sure I knew what had happened. At the swimming playdate that day, Mabel had probably received several painful bites from stinging flies. Now she was living in constant dread of them.

The owners acknowledged that this might be the case. The new, phobic Mabel had become highly reactive to flies. She appeared to be constantly scanning the environment. Inevitably, an occasional fly would get into the house, and Mabel would throw herself around in a state of panic, sometimes burrowing underneath things, just as she had done in the cab of the truck on the day of the triggering incident.

It wasn't a brave new world, it was a fearful new one. The unbrave new Mabel even panicked when simply loaded into the truck and burrowed under whatever clothes and blankets she could find. Mabel's doggy friend, the one she was playing with at the beach, became associated with the frightful event and, at the doggy friend's house, where Mabel used to be happy and carefree, she now hid and refused to play.

Whatever had happened to Mabel, it was a life-changing experience, and she now had a specific phobia—an excessive and irrational fear that disrupted normal functioning in daily life.

To treat Mabel, I made three recommendations. First, the obvious: Mabel's owners needed to avoid exposing Mabel to biting flies as far as that was possible. They also needed to enrich her home environment, providing plenty of distractions from her fears and entertainment to occupy her mind. Finally, we spoke about training her to go to a safe room, a place where she could get away from her winged nemeses.

I suggested using fly spray on her coat, a practice sometimes used on livestock. Mabel's owners were worried that this would be too toxic and declined to follow this suggestion. So, to facilitate a speedy recovery, I tried treating Mabel with an antianxiety medication.

A week later, Mabel had improved markedly. Then, out of the blue, she had a meltdown day. I switched to another antianxiety medication and, once again, she seemed much better, though she still occasionally scanned the room for flies.

On the down side, Mabel's owners felt that she was depressed. She spent more time inside the house than she used to, refusing to go out during peak fly times of the day. By the end of August, with the fly season winding down, Mabel was much better and I was able to taper off the medication.

The following spring, things began to unravel again. Mabel became much more nervous in April, as if anticipating the coming plague. By June she was not willing to go out of the house at all. At this point, her fear extended to all flying insects, including mosquitoes and moths. The situation had gone from bad to worse.

I again started her on the antianxiety medication, and this time added a long-acting beta-blocker to blunt the effects of norepinephrine, a brain-stress chemical that would be fueling her fear. The combination worked brilliantly, which was good, because her owners found her affliction so distressing that they had been considering euthanasia.

My own sister, Angela, had an insect phobia as a youngster and well into her teens. She was absolutely terrified of yellow jackets, having once been stung by one. The actual presence of a yellow jacket in Angela's same airspace triggered what looked like a panic attack. She became absolutely hysterical, screaming, crying, running away, hiding behind one of our parents or leaving the room. The very mention of the word "wasp," as we called them in our home of England, caused Angela to hyperventilate. So intense was her fear that even black-and-yellow sweaters would cause her to tremble.

When our family vacationed on the Norfolk Broads in East Anglia, on the east coast of Britain, we had rented a large six-berth cruiser to explore the extensive inland collection of connected rivers, canals, and large lakes of the Norfolk Broads. We sat in the galley eating breakfast when a yellow jacket suddenly appeared above the table.

My sister started to panic. Using his most authoritative voice, my father said, "Angela, there is no need to panic. Sit absolutely still and he will not hurt you at all."

Terrified, her pupils dilated in one of the human body's physical expressions of fear, Angela sat there sobbing. The yellow jacket circled around and eventually came to land on her arm. She froze. Then it stung her. She screamed in pain and became even more hysterical. Unfortunately, her yellow-jacket phobia was reinforced for another ten years or so. Thanks a lot, Dad!

Angela did not have the benefit of medication to expedite her recovery, but she and Mabel had a lot in common. They had a specific phobia. Specific phobias—a catchall rubric for a constellation of afflictions—have close parallels in people and animals. Natural fears are provoked by triggers such as spiders, snakes, mice. People can develop excessive, irrational fears—phobias—regarding situations, like riding in elevators or flying. Mabel's and my sister's phobias would fall into the "animal-type" of specific phobia, a fear triggered by animals, including insects.

I've also encountered agoraphobia in pets, as mentioned earlier when discussing separation anxiety, as well as natural environmental-specific phobias, another subclass, which are cued by events that occur in nature such as storms, heights, or water.

The prototypical natural environment phobia in dogs is weather phobia, sometimes referred to as thunderstorm phobia. Genetic factors may cause certain breeds of dogs to be more susceptible to developing storm phobia, with herding breeds overrepresented in the demographics. Little dogs, such as terriers and those in the toy breed group, are affected much less frequently.

In humans, too, genetics plays a role in the development of phobias. This has been evidenced in identical twin studies. Identical human twins raised separately can have identical phobias, such as fear of heights.

An associate who once worked with me was terrified driving over bridges. She would lean into the center of the car for fear of falling off the side. Apparently, her mother had an identical phobia. Of course, she could have learned this fear from her mother, but genetics may also have played a role.

Genetics aside, a key factor in the development of thunderstorm phobia in dogs may be nurture, in the form of a lack of acclimation to noise cues during critical periods of development. If a sensitive pup with a nervous predisposition were exposed to a terrifying thunderstorm during a critical period of development, and especially if the puppy was with a person who also reacted with terror at such turbulent weather, that pup might learn to dread thunderstorms from that point on.

The truth is that no one really knows what triggers storm phobia in susceptible dogs. Some behaviorists have suggested that poor hearing is a factor, and that the abrupt sound of the storm surprises and frightens them. Other theorists posited supersensitive hearing, and that the sound of thunder is almost painful to dogs. Neither of these contradictory explanations seems to hold water.

Whatever the cause of thunderstorm phobia, the dog's reactions show them to be terrified. Some seek out the comforting presence of their owner for support, clinging until the storm passes. Others hide under something or behind something, trying to escape. In extreme cases, when owners are not around, storm-phobic dogs will make desperate attempts to escape from their confines, breaking through screened or glass windows—anything to escape. In their panic they seem to be attempting the impossible, to outdistance the thunderstorm. Some dogs have run miles away from home.

Some dogs aren't merely triggered by thunder and lightning. By association, they may become terrified by darkening skies, high winds, heavy rainfall, and other elements of the storm. So the term thunderstorm phobia has been dropped in favor of the simpler storm phobia. Some researchers believe that falling barometric pressure or changes in the static electric field may be triggers. Ground vibrations or ozone in the air may also forewarn some storm-phobic dogs of approaching storms, making them nervous. Certainly dogs seem to be able to pick up on storms well before their owners can, sometimes when there is not a cloud in the sky.

I am in favor of the static electricity theory myself. Behavioral studies have determined that 50 percent of storm-phobic dogs find refuge under a sink or in a bathtub. Sinks and bathtubs act as electrical grounds, preventing the buildup of static electricity on the dog's coat. Dogs who get storm phobia are typically large dogs with thick pads and long woolly or thick or double coats, the animals most eligible for accumulating static charge.

If their bodies are not electrically grounded, pets can get static shocks during storms. This adds insult to injury—a painful zinger to an already dreadful situation. That would explain not only why they take refuge under sinks and in bathtubs, but also why the dogs with mild storm fear suddenly develop more severe reactions in midlife. They've probably been shocked by static electricity, and the effect was enough to push them over the edge into a full-blown phobia.

Some storm-phobic dogs also become generally sound phobic, frightened of sonic booms made by high-flying aircraft, quarry blasting, or even the sound of rumbling trucks. Others do not develop such ancillary fears. In fact, storm phobia and noise phobia are often distinct entities.

Many storm-phobic dogs are generally fearful and have other fear-related conditions, too. Quite often we see storm-phobic dogs who also have separation anxiety. In fact, at times storm phobia can be misdiagnosed as separation anxiety. A storm-phobic dog may indeed destroy property in his attempt to escape when his owner is away, a classic sign of separation anxiety. But the dog does so only occasionally, and only when the weather is extreme, which means he's more likely got a storm phobia. Dogs with separation anxiety panic and act out *every time* their owners leave.

Because the general weather patterns during storms are so multidimensional, desensitization to storm phobia is very difficult, bordering on impossible. If you own a storm-phobic dog, we usually recommend you set up a windowless or curtained safe place in the house for the dog to go when storms are predicted. Ideally, that would be a basement in which exposure to the sight and sounds of storms is limited. Think of the safe place as similar to a tornado bunker for families living in tornado alley. It's a great place to get out of harm's way.

Of course, you have to train the dog to go to the safe place when a storm is imminent. You should in fact accompany your dog down to this well-appointed location, demonstrating its safety. Equip the safe place with food, water, a crate with its door left open, toys, and food treats. When your dog goes into the safe place, you should give him a lot of positive attention and praise, and repeat this training over time, so the dog learns to take himself off to the safe place.

At Tufts, we also advise that you use white noise in the safe place to mask any distant rumblings of thunder. Also, you might keep the lights on. Bright-white lights can conceal any traces of lightning flashes that the dog might notice if the windows aren't completely shielded.

Various capes or wraps such as the Storm Defender and Anxiety Wrap have also proven to be helpful. The Storm Defender, with its flexible metallic-type lining, is principally an antistatic jacket. But its

Velcro straps can be applied snuggly to produce a squeeze-machine-like effect. In a study we conducted with this jacket, we found the Storm Defender effective in reducing signs of thunderstorm phobia by 70 percent. The Anxiety Wrap, which works by exerting controlled pressure on the dog's torso, was slightly less effective, but also produced a useful degree of improvement.

Some scientists believe that any improvement dogs show with thunder vests and T-shirts is due to the placebo effect. You might think that a placebo effect wouldn't work with animals, since they are not susceptible to suggestion. But it's not the animal who thinks it should work, it's the owner! If you tell owners that a pressure vest may calm their dog, around 30–40 percent will report that precise result.

Firework phobia is similar to storm phobia in some respects, but it is much easier to treat. Fireworks only have two predictable components, what the dog can see and what the dog can hear. Preventing a firework-phobic dog from seeing massive starbursts in the sky is relatively straightforward. You simply shelter him indoors in a shuttered room.

Dealing with the constant crackles and booms is not quite so simple. That's where desensitization comes in. Desensitization to the sounds of fireworks is eminently possible. You can play recordings of fireworks sounds, starting at low volumes and then increasing the level. At intervals, and as long as your dog remains calm, you reward her with praise and food treats. In time, your dog will learn that a fireworks boom in the background means that a freeze-dried liver treat is imminent.

Desensitization is a common treatment of specific phobias in people as well as dogs. Therapists reintroduce the trigger stimulus slowly, in increasing levels of intensity, until the desired level of tolerance is reached. Desensitization is usually coupled with counterconditioning, in which a person learns to associate the formerly phobic stimulus with more positive outcomes. Judging from the results we've seen with storm-phobic dogs, some kind of pressure vest could be helpful in the treatment of human phobias, too.

Another subclass of specific phobia in humans is the blood/injection/injury type. Fear of needles is common in children and holds on through adulthood with many people. Pets, cats especially, can develop injection-type phobias after a few visits to the veterinary clinic. A particularly painful blood draw or injection from a veterinarian might cue the animal to develop a white-coat phobia.

The good news here is that this phobia does not typically occur with any great frequency. It can be suppressed using situational antianxiety drugs and careful handling.

When still quite young, my daughter Keisha became terrified of needles after an inept nurse made a traumatic attempt to locate a vein. After that incident, she felt the hallmarks of panic every time she needed an injection. Her chest heaved, she felt faint and became certain she would pass out. Because some vaccinations are legally mandated and others are just plain necessary, she steeled her way through this situation, much to her credit. After many years of needle phobia, she has finally pretty much managed to overcome her formerly debilitating condition.

Another subtype of specific phobia is the so-called "situational type." In humans, as alluded to before, this might involve a fear of flying or a phobia connected to tunnels, bridges, elevators, public transportation, driving, or enclosed spaces. Extreme fears of this nature also occur in dogs and cats. Car travel might be a trigger for the afflicted pet, or being shut in a confined space, such as a crate or carrier. Desensitization and giving the cat or dog treats in the crate may lessen this fear in time.

The final subtype of specific phobia is mundanely termed "other type." Just as some children develop phobias to specific loud sounds or to costumed characters, so, too, can pets. By costumed characters, we really mean clowns. The latter condition even has its own name, coulrophobia. Many dogs are not comfortable in the presence of a person dressed in clown gear. And certainly many dogs are phobic of certain people who are in some way different in terms of the way they dress, physical appearance, affect, or gait. Halloween is not their favorite time!

Agoraphobia is a more complex phobia, defined as "fear of experiencing a panic attack in a place or situation from which escape may be difficult and where help is not available." For people, agoraphobia is frequently associated with panic disorder. Indeed, it is sometimes defined as a subset of panic disorder. In severe cases, people become unable to leave their homes or safe havens. They are afraid of open spaces or, more basically, of being outside.

One of the most shocking and memorable cases of agoraphobia that I came across in animals involved a cow. Cassie, an otherwise healthy-looking 1,500-pound black-and-white Friesian, was confined to a roomy, well-bedded stall in an animal sanctuary in Hopkinton, Massachusetts. I was called in because Cassie would never leave her stall. In fact, she could not be induced to take a single step outside her stall, even when tempted with her favorite treat, bread. This poor cow was frightened of her natural environment!

Looking into the Cassie's history, I realized why she had developed this incredible phobia. A few months before, Cassie found herself at a slaughterhouse in Hopkinton, minutes away from meeting her maker. As the doomed beasts intended for slaughter mooed and bellowed their way up a ramp to their execution, Cassie rebelled. Clearly, she did not feel ready to shuffle off this mortal coil.

Instead, exhibiting the strength of ten cows, Cassie somehow clambered over a six-foot-high barrier, bursting out of the slaughterhouse with the power and agility of a puissance horse, to find herself in the main street of pleasant little Hopkinton, cars and trucks whizzing by her. The brave cow continued to flee, galloping down the main street of Hopkinton as if to start the Boston Marathon.

The escaped beast was eventually cornered by the authorities and loaded into a stock truck. The animal control officers had a tough decision to make. Would they give Cassie a reprieve or take her back to the slaughterhouse? The case of Cassie, the cow who did not want to die, became something of a local cause célèbre. Something about Cassie's desperate escape to freedom plucked at the heartstrings of the public.

That's when the good folk at Maple Farm Sanctuary stepped up and said, "We'll take her." Maple Farm Sanctuary's decision was very popular. The reprieved cow quickly adapted to her new home. In the sanctuary of her stall, she was safe, comfortable, and well fed. She did not want to face the horrors and panic of the outside world again. There are slaughterhouses out there! So Cassie simply stayed put.

My job was to coax her out. Her new keepers wanted her to taste the joys of freedom and experience the wonderful meadows that lay behind the barn. I'm afraid the treatment I suggested was not supersuccessful. It did help a bit. I had her eating bread, her favorite ration, and taking several paces away from the stall to obtain successive mouthfuls. This worked quite well, especially when facilitated by a mild tranquilizer. Attempts to move her farther than a few paces met with rigid resistance, with the stubborn cow channeling her inner mule. Eventually, the sanctuary owners decided that if that is the way Cassie preferred it, she could live out her days in the stall.

I have seen some pitiful specimens of dogs suffering from agoraphobia. Many have come from abusive situations and their behavior resembles Cassie's—they prefer the safety of their enclosures to any freedom their new owners might offer.

Canine agoraphobia often involves balking. By balking, I mean refusing to venture out onto city streets. This affliction is most common in urban pets, just as agoraphobic mental disorders are most frequently seen in urban-based humans. The unnatural city environment—the concrete jungle, so to speak—seems to present problems all around. The balking dogs I have seen have all come from Boston or New York City.

Another agoraphobia case I saw involved a horse that was boarded at a barn a bit north of our veterinary school in central Massachusetts. One winter's day, as the horse was being led from its stall to an outdoor paddock, a massive sheet of frozen snow rumbled across the roof and fell to the ground below, smashing noisily into a thousand different pieces.

The horse spooked and abruptly reared up, but was then controlled by his handler and led on to its paddock. The problem came later, when the horse balked at returning to its stall. For weeks, he remained in the lower paddock, refusing to be led back to the barn, showing signs of panic that intensified as he got closer to the feared location.

Although an argument could be made that this horse was expressing a specific phobia, the case also met some of the criteria for agoraphobia. The animal perceived returning to the stall environment as threatening, panic-inducing, and difficult to escape from. Diagnosis can be a moot proposition to a suffering animal. The horse was in the grip of a phobia, and whichever one was irrelevant to the animal. The treatment involved gradually reintroducing the horse to the barn, facilitated by the use of the antianxiety medication buspirone.

Treatment of agoraphobia can be very difficult and not guaranteed to work, whether with cows, dogs, or people. For people, cognitive behavioral approaches, like exposure treatment and relaxation techniques, can limit symptoms of anxiety and panic, but most often pharmaceutical treatments are needed as well. Agoraphobia comprises about 60 percent of all phobias and is such a stubborn problem that it is sometimes classed as a medical, rather than a psychological, condition.

It is one of the few anxiety-based conditions that's so severe it receives specific medical labeling.

Although I have not actually crunched the numbers, my suspicion is that "specific phobias" would be more common than agoraphobia in nonhuman animals. They certainly occur frequently, and they can be seriously debilitating and highly distressing for both the animal and the owner. Persistence in calm, kind desensitization combined with anti-anxiety medications can sometimes help the phobic pet, but treatment is a challenge for creatures great and small.

CHAPTER TWELVE

Senile Dogs, Cats . . . and Cheetahs

Dementia and Alzheimer's Disease

Whoever saves a life, saves the world entire.

—THE TALMUD

Marty, a fifteen-year-old neutered male Border collie, had been taken to his local veterinarian because he no longer seemed to recognize familiar people. He had less and less interaction with his owner, who noticed him staring off into space quite a bit, gazing at blank walls as though he might see something there.

There were other disturbing quirks. Marty had taken to pacing aimlessly from early evening into the wee hours. Although his sleeping at night was fitful, he slept most of the day. In addition, though fully housetrained for all his life, he was suddenly having accidents in the house.

Medically, there was nothing amiss. Marty checked out just fine in all areas. But the vet who referred Marty to me correctly diagnosed canine cognitive dysfunction (CCD), otherwise known as canine Alzheimer's disease.

The acronym for the signs of this condition is DISHA:

- *D* for *Disorientation*
- *I* for altered social *Interactions*
- *S* for *Sleep* disturbance
- *H* for *House soiling*
- *A* for altered *Activity* level

Poor old Marty had the behavioral full house. Knowing a fair amount about CCD, the local veterinarian put Marty on a medication, selegiline, which can sometimes reverse the clinical signs, although it cannot cure the problem, any more than the medicines prescribed for humans with Alzheimer's can cure the disease. But it can buy affected dogs more quality time.

Selegiline is an old-fashioned monoamine oxidase inhibitor (MAOI), an antidepressant that blocks the breakdown of dopamine in the central nervous system. Dopamine has effects on consciousness and mood. Alzheimer's syndrome may involve a lack of dopamine. Increasing dopamine levels in the central nervous system should thus help reverse at least some signs of the disease. This is what can and does happen for some dogs.

With selegiline treatment, about a third of dogs with CCD have their condition almost completely reversed. Another third show noticeable improvement. The remaining third show no improvement at all, perhaps because the disease has progressed too far or their dementia has other sources. Unfortunately, because the disease is progressive, the improvement only lasts for a few months.

One amazing success story I heard during premarketing trials for selegiline was about an eleven-year-old Afghan hound. He was treated with a veterinary version of selegiline, trade name Anipryl. Selegiline is used to treat human Alzheimer's under the medical trade name Eldepryl.

After treatment with the drug, the Afghan was behaving like a two-year-old, his owners reported.

"That's great," the researcher said.

"No, it's not," the owner replied, not entirely seriously. "We didn't like him when he was two. Would you please reduce the dose to make him act like he is six?"

After six weeks on selegiline, Marty was much improved. His house soiling and aimless pacing decreased considerably. Two and a half months after beginning the selegiline treatment, there was some backsliding, so Marty's owner brought him to see me to ask if anything more could be done.

In my consulting room, the old Border collie looked forlorn. His head was bowed. He had a vacant expression and appeared barely interested in his new surroundings. He did not investigate the premises, sniffing out novel smells the way a normal dog might do. Yet this afflicted creature slowly made his way across the room to approach me. I was touched and flattered when he wagged his tail slowly and offered his head to be petted.

Marty's owner wanted to know if he was salvageable or whether the best thing for him would be to put him to sleep. I did not think it was time yet to go that route.

"Look at him," I said, gesturing to the dog at my feet. "Marty still enjoys many aspects of life, doesn't he? You say he's eating well. He obviously enjoys petting and attention. Before we throw up our hands, I think there is more we can do to try to reverse symptoms of his condition."

Hearing this, the owner almost wept for joy. I just hoped I hadn't overstated my case.

I switched Marty's diet to an antioxidant-rich dog food, Hill's Prescription Diet b/d, a proprietary formula designed to address issues with brain health and aging. In research studies, aged rats were fed blueberries, rich in antioxidant flavonoids and vitamin C, and seemed to gain a new lease on life. Antioxidants can help turn things around with canine Alzheimer's, too, or at least slow its progression.

I recommended two dietary supplements for Marty, acetyl L-carnitine and coenzyme Q10 (CoQ10). The first is an antioxidant that has other neuroprotective effects. CoQ10 is an antioxidant, too, and is known to boost cellular energy production. Finally, I prescribed the sleep hormone, melatonin, to be given to Marty at night to help him sleep more soundly. Melatonin also happens to have antioxidant properties and has been shown to increase life span in experimental animals.

I asked Marty's owner to keep his life interesting and engaging, giving him as many walks as he could handle, play dates with other dogs, and plenty of new toys. In research studies, novel environments have been shown to give aged rats new vigor. A rich social life has been shown to help stave off Alzheimer's disease in people.

Within a few weeks under this new regimen, Marty was moving around more and showing more interest in his surroundings. When his owner went to give him a boost into the car, he surprised her by jumping in by himself. He slept through the night, and was almost "accident-free" in the house. Clearly his condition had improved.

When she first brought Marty to see me, she had said that, when she looked into Marty's eyes, she saw an unfocused, anxious creature gazing back. "Now, he's back, and it's a beautiful thing," she told me. "It was sort of like there was no one home before. Now the lights are on and Marty is in."

Even though Marty's condition would deteriorate in the future, we had bought him some more quality time. According to the folk wisdom formula, one more year in his life would equal about seven years in ours. The one-to-seven ratio has a lot of qualifications attached to it, but it's a useful rule of thumb. So when you give your pet a three-minute massage, credit yourself with a full twenty minutes.

Older cats get a feline version of Alzheimer's disease, called feline cognitive dysfunction (FCD). Typically, FCD develops in susceptible cats above eleven or twelve years of age. The onset is insidious and almost unnoticeable. The first sign of the condition might be a loss of interest in play and excessive daytime sleeping. Then comes the aimless walking and disorientation.

Kitty will go from bad to worse. The disease may progress to staring blankly at walls, decreased interest in food, and house soiling. As with canine cognitive dysfunction, nighttime hours seem the worst. Affected cats pace the floor as if they're lovesick, caterwauling and moaning in a haunting and pitiful way.

I bore personal witness to a case of FCD in Monkey, my seventeen-year-old cat. With Monkey, the first sign was a changed personality. She became much less playful and interactive. *She's old*, I would think. *It's only to be expected.* Yet her mom, Cinder, did not undergo the same personality change, even though she was a couple of years older.

As time went by, Monkey started to have accidents outside the litter box—right next to the litter box instead of in it. I tried everything I knew to get her back to using the box: added other litter boxes in the house, tried different litters, scooped the box more frequently, all the tricks that usually work so well, to no avail. Luckily, I was feeding her a high-fiber ration so the stool she produced was compact and easy to clean up.

The next stage of Monkey's decline involved constant nighttime perambulations. She accompanied these walks by moaning as if she were in pain. She had formerly yowled like that, but only when attempting to throw up a hairball. Now it was just ghostly crying in the middle of the night. By then, my wife and I knew our cat was in real trouble.

When Monkey reached nineteen years of age, on her last day on this planet, she wandered aimlessly across the living room floor, showed no interest in her food, and then disappeared. It took us ages to find her. We searched the house all day and finally had to pack it in for the night. As we were getting ready for bed, we heard a faint moan coming from the bathroom. Peering through an opening in the marble-tiled tub surrounds, we saw Monkey at the far end, hunkered down and almost out of sight.

We had to use a hammer and pry bar to break apart the tiles in order to reach her. Sadly, it was too late to do much. She was clearly in the last stages of life. Even before we could humanely dispatch her to prevent further suffering, Monkey passed away in our arms. Though we knew she was unwell, we were still devastated.

Human Alzheimer's is hauntingly similarly to the dog and cat version, or perhaps it is more proper to say it is the other way around. Unfortunately, I have had firsthand experience of this condition, too, as my mother died of complications related to Alzheimer's disease. As with the animals, the beginning of her decline was insidious. She would forget the odd word, like my daughter Victoria's name, referring to her as "your little girl."

Early on, my sister Angela informed me that our mom's behavior indicated the initial stages of dementia, but I remained unconvinced. I suppose I wishfully ascribed my mother's forgetfulness to senior moments. Then I greeted her at the airport after she returned from a transatlantic flight.

"I was tucked into bed by the nice nurses," she told me. She wasn't being fanciful. She really did believe the flight attendants were medical personnel. Only then did I realize how much she had deteriorated.

During that visit I had her stay in my guest room. I showed her the bathroom, right across from her room. That first night we woke to hear a tinkling noise in the bathroom off our own bedroom. I found Mom in there, confused and apologetic. I escorted her back to her own bedroom. I discovered a trail of urine from her room at the far end of the house all the way to our bedroom.

I felt so sorry for her being disoriented—which would be the *D* for disorientation in the DISHA acronym. That same trip she turned on the gas stove and forgot to light it. She also put the kettle on without turning on the gas, which was far less dangerous.

She continued to decline when she returned home to England, her condition going from bad to worse, as it always does with Alzheimer's disease. One by one she gradually ticked off all the elements of the DISHA acronym. I was there with her the day the diagnosis of Alzheimer's disease was confirmed. Afterward, walking down the hospital corridor hand in hand, she told me that she had a wonderful son now living in the United States.

"It's me, Mom," I said.

"Oh, I am bad, aren't I?" she replied.

Mom eventually faded further and further away. Her activity level fell to near zero (note the *A* in the DISHA acronym, for activity). She slept each day away, and was put to bed at six p.m. At this stage, I was not sure she even recognized me.

On my very last trip to see her, a few weeks before she died, I gave her a big hug as I was leaving the home. I made my way toward the door saying, "I have to leave now, Mom. Time to catch my plane."

She straightened up from her hunkered-over posture, looked mistily at me, and gave me a weak wave. As I looked into her eyes, I saw a tear roll down her cheek. The light was on and someone was home after all. It broke my heart.

This was the woman who had first set me on my path in life. Gwen Dodman taught me the fundamental basis of One Medicine; that all life is related and that animals are more like us than we sometimes think.

Whether it occurs in people or in pets, Alzheimer's is a cruel disease.

Piece by piece, it robs the afflicted of their memories and leaves them an empty shell of what they used to be. Some say it's worse for the family, who have to witness the inexorable deterioration that occurs with this merciless condition. There may be something to that. We don't know how the afflicted feel, since they can't tell us. Alzheimer's seems to take people back to childhood, erasing memories in reverse. Living with Alzheimer's is a *Groundhog Day*–like existence. Each day there is a mental whitewashing of what came before.

At some point, people who are affected with the disease do not even realize they have it. As my colleague Lou Shuster says, "If you think you may have Alzheimer's, then you don't."

The steady, unstoppable course of Alzheimer's over time is a feature of the disease in both animals and people. In dogs, various psycho-motor tests have been used to examine and monitor the condition's progression. Demographically, the number of dogs affected with CCD increases steadily from age ten onward. Similarly, human Alzheimer's disease is more prevalent in older age groups.

Another parallel between human Alzheimer's and animal Alzheimer's is the pathological change that takes place within the brain. In both humans and animals, a protein called beta-amyloid accumulates between brain cells, developing into what are called senile plaques. Amyloid accumulation is associated with a decrease in brain volume. Such buildup of plaque correlates closely with the degree of cognitive impairment in both people with Alzheimer's disease and pets with cognitive dysfunction.

So-called neurofibrillary tangles of a protein called tau also correlates with the cognitive impairment of Alzheimer's. Though tau tangles do occur in aged dogs, the link between this change and clinical deterioration is not quite so clear as it is in humans.

Studies have ascertained the link in another species, a surprising one: cheetahs, *Acinonyx jubatus*, the fastest land animal in the world. Researchers at the School of Veterinary Medicine of Azabu University in Japan have demonstrated that in cheetah research subjects, spontaneous development of amyloid plaques and tau tangles do occur. Both are seen more frequently in aged animals. Similar to human patients with Alzheimer's disease, in cheetahs the amyloid changes were mostly distributed in the parahippocampal cortex in the CA1 region. Two cheetahs in the study with the most severe cognitive impairment also had the most severe tau pathology.

Compelling pathological similarities have made canine cognitive dysfunction an interesting model for study of the human condition. For example, treatments suspected to be effective in people can be tested much more quickly and easily in dogs. A constellation of genes have been implicated as increasing risk for Alzheimer's disease in people. To study genetic contributions to CCD in dogs might help further elucidate the role of individual suspect genes and their most relevant interactions.

One relatively new theory posits that Alzheimer's disease may be a form of diabetes. Type 3 diabetes is what some physicians are calling it. Brain cells need glucose to function. Insulin packs glucose into cells. If brain cells are bombarded with too much insulin for too long for reasons of a bad diet, they become insulin-resistant. When this happens in the brain, watch out. Certainly the incidences of type 2 diabetes and Alzheimer's disease have increased in lockstep of late. I believe the type 3 diabetes hypothesis could be studied more rapidly and easily in canines than in people.

Another treatment that is used for people with Alzheimer's is Namenda, a drug with the generic name of memantine. Namenda is an NMDA blocker, prescribed off-label to alleviate obsessive-compulsive disorders in animals and people by negating the effect of the excitatory neurotransmitter, glutamate. Another theory about the causes of Alzheimer's disease is an overactive glutamate system. It's thought that beta-amyloid does its damage by wreaking havoc at the glutamate-sensitive synapses. High levels of glutamate cause calcium to flood into the cells, which triggers overproduction of toxic substances, including tangled tau proteins. This can lead to cell death. Namenda helps regulate this and may slow the progression of cognitive deterioration in dogs with CCD or cats with FCD, as it does in people.

Nitromemantine, a derivative of memantine, is also designed to block glutamate receptors and to reestablish damaged synapses. In a mouse model of Alzheimer's, the treatment reversed synapse damage almost completely in a matter of months. Nitromemantine might be a more effective drug than memantine, as the latter is repelled from the so-called eNMDA site by a conflicting electrical charge. Unfortunately, this drug is not yet available for routine use.

Another promising drug, Aricept, works by allowing the buildup of a neurotransmitter called acetylcholine in the gaps between certain brain cells. Acetylcholine is a neurotransmitter that can be at a low ebb in patients with Alzheimer's. I feel sure this medication would help animals suffering from dementia, but it has yet to be tried in nonhuman applications.

Along the same lines, the herbal derivative, huperzine, which I mentioned for treating complex partial seizures, is known by herbalists to be a treatment for cognitive impairment. One other potentially useful pharmacologic action of huperzine is that it binds, however weakly, to NMDA receptors, so it could conceivably act like Namenda and Aricept together as an NMDA blocker and acetylcholine enhancer, a double whammy of cognitive boosters. While huperzine's value in treating CCD can be posited from first principles, there is not yet any direct clinical evidence of its efficacy in people or animals. With that said, I would expect it to be helpful in dogs and cats with cognitive dysfunction—and cheetahs, too, for that matter.

A number of new experimental treatments for Alzheimer's have come along recently, many developed first with animal models. One of the more promising treatments is with a drug which currently only has a generic name, bexarotene, tested at Case Western Reserve University. It has been shown to clear more than half the amyloid plaques in a mouse within seventy-two hours of administration. Ultimately, the total reduction of amyloid was 75 percent. Researchers claim that bexarotene "programs the brain's immune cells to eat the amyloid deposits." This drug may eventually find its way into human clinical trials, although not all researchers are enthusiastic about it, because other studies have not been able to replicate the earlier promising results in mice. Science marches forward, but sometimes it takes a step back along the way.

Canine and human brains both show similar structural changes during the aging process. One effective way of studying this change is by using voxel-based morphometry, or VBM, a type of MRI technique that I mentioned before in the discussion of compulsive disorders. With VBM, researchers found frontal lobe atrophy and age-related changes in areas of the cerebral cortex. Also affected were the thalamus, one of the brain's relay stations; the cerebellum, which affects motor control, some cognitive processes, attention and language; and the brain stem, responsible for controlling vital functions. There is also atrophy of the brain and enlargement of the normally fluid-filled spaces within the brain, seen in both the normal course of aging and more severely in Alzheimer's in humans and dogs.

It is now possible in people, and therefore in dogs and cats, to scan the brain for beta-amyloid. We can detect an early increase in beta-amyloid level before clinical signs appear. This new advance lends itself to early diagnosis and preemptive treatment, although the effectiveness of preventive treatments in clinically normal patients is controversial. I would advocate that we study the efficacy of the new treatments in at-risk dogs instead of waiting for the full-blown condition to develop. While not all studies of treatment in a mouse have translated into efficacy in human trials, I feel sure that the canine condition has more to offer.

I often stop in the middle of such meditations to give a thought to my late mother. She would be delighted that research in pets might benefit humans. Gwen Dodman gets my nomination as a patron saint of One Medicine. Considering the cruel way she passed away, losing her faculties to the ravages of dementia, there is poetic justice in applying her belief in the unity of all life to the battle against the disease that killed her.

The number of people with Alzheimer's increases each year. An effective treatment is tragically overdue. A similar increase in the prevalence of cognitive dysfunction is also being seen in the population of aging dogs and cats. As we care for them more effectively and provide them with healthier, better-balanced diets, our pets live longer. That's a good thing. But some live long enough to develop cognitive impairment. Canine cognitive dysfunction, like Alzheimer's disease, is a bad moon on the rise. We desperately need to find a way of treating it.

CHAPTER THIRTEEN

The Beagle with ADHD

Behavioral Problems with Medical Roots

I have been told I have ADHD but I don't think I—hey, look, a squirrel!

—ANONYMOUS

The owner of a nine-year-old spayed female beagle named Emma described her as "hyperexcitable" and "hyperactive." Emma also showed some signs of thunderstorm phobia and was obsessed with food. I was not very concerned about this last aspect of the dog's behavior. Beagles are notorious food-oriented "chowhounds." Additionally, Emma also exhibited signs of separation anxiety. She would become anxious whenever she noticed the owner was leaving. Emma would howl as her owner headed off down the front walk. She destroyed things or had accidents in the house when alone, and delivered the classic squat-and-pee greeting response upon the owner's return.

As my primary diagnosis, I jotted down "separation anxiety." Also on the list was generalized anxiety and, somewhat dubiously, possible obsessive-compulsive disorder involving food. But in the back of my mind I considered the remote possibility of attention deficit hyperactivity disorder, better known under the rubric of ADHD.

I decided to treat the separation anxiety first, using a training program that fostered independence, environmental enrichment and medication with an antidepressant, Clomicalm. The typical time scale for improvement with such medication is two months. At that point, Emma's owner reported that her dog was considerably better. The beagle still displayed some signs of separation anxiety, however, was still excitable, particularly in the car, and was still obsessed with food.

But we had made some progress. I was moderately happy with this report. I allowed the situation to drift for a few more months before inviting the owner to bring Emma back for a follow-up appointment.

"How much better is Emma's separation anxiety?" I asked. "Give me a percentage of improvement."

"I'd say Emma is about thirty percent better," she replied cheerfully.

I swallowed hard, disappointed. "Thirty percent, is that all?"

"Well, maybe fifty percent." I saw that the woman was simply hoping to make me feel better by quoting a higher percentage.

Even stretching the improvement to 50 percent, Emma's progress was unsatisfactory. The conversation turned once again to her unruly behavior. The owner reported that Emma was positively "ballistic" in the house, tearing around the place and practically ricocheting off the walls. In addition, she was still totally unmanageable in the car.

The owner kept invoking the word "hyperactive" to describe Emma's behavior. She was doing my work for me. After all, what does H in ADHD stand for?

"Okay," I finally said. "Let's do a trial to rule out ADHD."

I arranged for the beagle to do a Ritalin trial at home. The results were jaw-dropping. Emma's owner reported that a ten milligram dose of Ritalin totally calmed the dog down. This was not a 30 or 50 percent change. This was a 90–100 percent improvement. The owner added that Emma did not appear sedated, but was acting "just like a normal dog." To her, the result was nothing short of a miracle.

Armed with this new information and a putative diagnosis of ADHD, I had the owner give Emma Ritalin twice daily to see if we could achieve sustained improvement. A week or so later she called back to say that Emma now tolerated thunderstorms, was calm in the car and her food obsession was gone. You might argue that Ritalin cured her food obsession because it reduced her appetite, but the other responses were difficult to explain unless Emma had ADHD.

The only remaining issue was some desultory hound-style howling when Emma was left alone. Everyone was happy with this result, including Emma. My conclusion was that ADHD was contributing to Emma's clinical signs and that most of her behavior problems were secondary to ADHD. A medical condition, in other words, caused the behavioral condition.

Over the years I've seen many dogs like Emma with an assortment of behavior problems that owners describe as overactivity or hyperactivity. Some have responded to treatment with stimulants such as Ritalin and methamphetamine. I've had my share of failures, too, in the response to test-dosing dogs with a stimulant.

Not every apparently hyperactive pet necessarily has ADHD. Some of these subjects may have been simply overactive. Being overactive is a behavioral state resulting from lack of exercise, an improper diet and a boring lifestyle. That is all too often the case with pets in modern suburban households with no outlet for their pent-up energies. I almost always address underlying lifestyle issues before prescribing Ritalin. In my experience, ADHD is uncommon and is something of a curveball diagnosis.

Many behavior issues have medical underpinnings. A listless cat might have developed thyroid problems. Nocturnal anxiety could have its roots in a serious illness such as cancer. Anatomical irregularities in certain breeds can lead to sleep apnea. Determining if a behavioral issue has a medical cause is often a vital step in my practice when presented with problem pets.

Dr. Samuel Corson at Ohio State University in the 1960s established the validity of a diagnosis of canine ADHD. His team's research on the effects of stress in dogs found that laboratory dogs exposed to random, mild electric shocks through the floor of their cages developed a hyperkinetic syndrome. Yet these dogs, including a dog called Jackson, who became aggressive as well as hyperactive, responded wonderfully to treatment with stimulants. The paradoxical calming effect of a stimulant drug was analogous to the response of hyperactive children to such treatment. A groundbreaking model for study of human hyperactivity had been created, out of which flowed the present-day treatments for ADHD.

In Pavlovian conditioning trials, dogs that had difficulty paying attention responded positively to treatment with stimulants. At about the same time, other researchers found hyperactivity in rodents and cats. It seemed that all animals, including humans, could suffer from this troublesome condition.

The veterinary behaviorist Dr. Andrew Luescher encountered ADHD quite frequently in certain breeds of dog, including working German shepherds. To confirm the diagnosis, if a dog appeared clinically hyperactive, Luescher would bring him to the veterinary hospital to make baseline measurements of behavior, heart rate, and respiratory rate. Then he administered a stimulant, such as amphetamine, and continued taking measurements over four hours.

In a normal dog, the expectation would be that the dog's activity level, heart rate, and respiratory rate would increase. Logically, the dog would be stimulated by a stimulant. In a dog with ADHD, by contrast, a paradoxical calming response would occur. The administration of the drug was accompanied by reductions in heart rate and respiratory rate.

The latter response, Luescher argued, supported a diagnosis of ADHD. The logic was somewhat switched around. If subjects responded to the treatment, Luescher theorized, then they must suffer from the syndrome.

Another veterinary behaviorist, Dr. Walter Burghardt, was so convinced by the concept of canine ADHD that he once said he diagnosed it in over 30 percent of his patients and treated it with Elavil, a tricyclic antidepressant with stimulant properties. In Burghardt's view, owners do not usually come to behavior clinics stating that their dog has ADHD, but have various other complaints about their dog's behavior. It is up to the clinician to determine ADHD as the root cause.

One added note regarding Dr. Corson: today, he is most remembered for pioneering the use of therapy dogs to help alleviate certain mental conditions in humans. One famous case involved a nineteen-year-old psychotic man who stayed in bed all day and did not respond to normal treatments. When a dog was brought into his room, however, the young man totally opened up. The mere presence of the pet was therapeutic. It's remarkable that a researcher who focused on a canine model for ADHD established that pets can serve as a calming influence for agitated people. Once again, it highlights the fact that the benefits of medical research can flow both ways, from humans to pets and from pets to humans.

Other medical problems that can generate behavioral changes are such dire conditions as liver failure, rabies, cancer, and a rare disorder called lissencephaly, or "smooth brain." The danger here, quite apart from those presented by the conditions themselves, is that a veterinarian might miss the underlying reasons for a pet's strange behavior.

When the liver fails to do its job for any reason, toxic substances accumulate, such as ammonia and fatty acids. Because of this, the brain is also affected and neurotransmissions get out of whack. The processes that occur in liver failure in humans and animals are similar, as are the behaviors and neurological signs that this dysfunction produces. Behavioral changes in dogs associated with this condition include loss of appetite, depression, and lethargy, as well such bizarre quirks as aimless wandering, head pressing, circling, pacing, and even compulsive eating and drinking. Some animals appear blind. Others develop partial or full-blown seizures, sometimes progressing to coma and death.

In extreme liver failure in an old dog, no veterinarian would confuse this serious medical condition with a behavioral one. In the early stages of liver failure, however, particularly in a youngster, the behavioral signs may not lead a veterinarian to suspect liver failure and immediately check for it.

A typical scenario might be a young dog who has a congenital condition, in which blood that would normally traverse from intestines to the liver goes directly into the circulation system instead, bypassing the liver's filtering and purification processes. This is called a portosystemic shunt and is quite common in some small breeds of dogs, particularly in Maltese terriers. Various behaviors, including unpredictable bouts of aggression, pacing and circling, and partial seizures may occur, all due to the underlying physical condition. In some affected dogs, behavioral issues are more pronounced after a high-protein meal. These behavioral signs form grounds for suspicion of a liver disorder.

Although portosystemic shunts are uncommon in human children, they do occur. In children, other liver diseases, such as failure of the bile duct development or hepatitis, produce similar signs, including personality and behavior changes, mood swings, impaired judgment, and seizures. Of course, a physical workup with blood testing and imaging can easily ascertain the liver failure in people and animals. So if you have a child who has bouts of "acting weird" at school, or a Maltese terrier who starts stargazing or acting aggressively after a meal, liver tests could be in order.

Rabies in humans and animals is caused by the same infectious agent, involves the same pathophysiologic processes, and results in similar signs in humans and animals. The virus travels up peripheral nerves to the brain and causes severely disruptive neurological symptoms. These include confusion, agitation, aggression, paranoia, terror, and hallucinations, usually progressing to death. Both people and animals may express the so-called furious form of rabies. Rage is the outward expression of the inner disruption of the brain. Paralysis occurs with or without hydrophobia. Basically, to know the disease in one species is to know it in another. Owing to the vaccination of dogs, cats, and people who work with animals, rabies in pets and people is extremely rare in the United States. Prophylactic treatment given immediately after a bite is also highly effective, and it is necessary since the untreated disease is 100 percent fatal.

Any cancer that causes abnormal growth of cells within the brain can affect its function. Since the growth occurs within what is an enclosed space, the skull, the expansion of the tumor causes an increase in pressure. The consequent disruption of normal brain activity is sometimes accompanied by partial or generalized seizures. Brain tumors are always something to watch out for in elderly pets when they start to behave oddly. They may display different-sized pupils or a pronounced head tilt. Or they might circle in one direction or exhibit seizures.

Some years ago, I missed a diagnosis of a brain tumor in a cat. The cat's owner, an elderly woman and a resident of Cape Cod in Massachusetts, complained that her cat was biting her hands and forearms. Because of ongoing damage to her delicate, aging skin, the owner dearly wanted these attacks to stop.

Working on the assumption that common things occur commonly, I treated this six-year-old feline as if it was a case of what I call the alpha cat syndrome. I advised the owner on how to avoid aggressive attacks, how to repel an attack if necessary, and how to teach her cat to earn valued resources. She reported that this treatment had helped in the short-term. A couple of months later the local vet called to tell me that the cat had started circling in one direction. The diagnosis: a brain tumor.

I should have realized that a six-year-old cat does not suddenly start exhibiting an odd behavior problem. If the problem were purely behavioral, the cat would have been showing this pushy, assertive behavior at a much earlier age.

A thirteen-year-old beagle-sheltie mix called Buddie whom I saw relatively recently suffered from what appeared to be periodic panic attacks since the age of ten. The episodes began with him leaping up as if he had been bitten. Following that, he would spend days in a state of agitation and terror with his family helpless to make him feel better. His owner reported that he could not be verbally interrupted when the attacks occurred and stated that a bout usually lasted four to five days. Although I initially considered panic attacks or noise phobia as possible causes for his condition, it occurred to me that they might also be brought on by partial seizures. Seizures in a dog older than ten years are frequently the result of a brain tumor. In Buddie's case, it would have had to be a slow-growing tumor because of his three-year history of "panic attacks." But that was still possible and would have explained the geriatric onset of the behavior and the sudden change in behavior in a dog with a previously even temperament. An MRI would have settled the diagnosis, but an MRI costs up to $2,000, and quite likely there would have been nothing that could be done, even if one were discovered. Buddie had already lived past the life expectancy for a dog of his size and breed mix, so I decided to treat him symptomatically and to forgo pursuing a definitive diagnosis. His owner agreed—but a brain tumor causing partial seizures was high on the list of possible causes for his condition.

For humans, brain tumors present signs similar to the ones we see in animals. Behaviorally, there can be confusion and personality changes, leading to altered behavior and sometimes aggression. Depending on where the tumor is, the signs can be slightly different. In general terms, they are equivalent from species to species. Treatments are the same, too, including surgery, radiation therapy, and chemotherapy. If you were a radiation oncologist working with people, your skills would easily transfer to treating domestic animals, although you would have to be prepared to take a serious pay cut. The opposite would also hold true, that a veterinary radiation oncologist would have a flying start in an oncology clinic.

Lissencephaly is a brain condition that occurs in both dogs and people. The brain surface, which normally is wrinkled like a walnut, is instead smooth like a billiard ball. Most affected dogs and people are seriously impaired in learning and behavior. Lissencephaly has been reported in breeds such as Lhasa apsos, Irish setters, samoyeds, and wirehaired fox terriers, which implies that faulty genetics may be involved. In humans, lissencephaly can be caused by viral infections during the first trimester of pregnancy, or insufficient blood supply to the fetal brain, also in early pregnancy. But genetic causes are known, including a mutation of the so-called relin gene.

Interestingly, the relin gene is named after a spontaneous mutation in the so-called relin mouse. Without this gene, mice have abnormal brain development and coordination problems. If the genetic causes of lissencephaly were studied in dogs, the same gene might be implicated.

Behavioral conditions with medical underpinnings can have heartbreaking repercussions in pets. Dogs that may simply be suffering from medical conditions are blamed for bad behavior, as if they were at fault. A cat that scratches its owner suddenly becomes less loved and is ultimately rejected. This blame game sometimes leads to euthanasia, when in reality a corrective operation or dose of medicine might solve the problem.

Pain, from any cause, will affect the behavior of animals and people. The receptors that sense pain are the same in animals and people, the nervous pathways that transmit pain signals are the same, and the brain centers involved in receiving pain signals are the same. It only makes sense that the outward expression of pain is pretty much the same.

Pain underlies many a behavior problem and animals respond to the same analgesic drugs that are employed in people for pain relief. We see reduction of both the outward and inward signs of pain. Specifically, opioids, aspirin, and other nonsteroidal painkillers all deliver relief in dogs, cats, and other animals. Topical and regional anesthetics when appropriate are also highly effective in controlling pain.

Pain can sometimes be subjective. Injured soldiers returning from war require less morphine to control their pain than soldiers with identical injuries on the way to the front. We humans increase the pain we feel by projecting what the injury will mean for us in the future. With dogs, there are probably no such projections. Pain for animals is appreciated in the moment.

Anxiety increases the sense of pain in people and animals. Drugs like nitrous oxide and morphine do not completely alleviate pain in people: they just make recipients care less about the pain so they can tolerate it better. Though I cannot prove it, I think the same dynamic might apply to animals.

I do know, however, that morphine works in dogs to control pain. At Tufts, we have done studies on pain relief in dogs using epidural morphine to control postsurgical pain. In another experiment, we used an opioid called butorphanol in dogs after limb amputation surgery. It worked well. After twenty years of experience as a veterinary anesthesiologist, I can say with confidence that animals experience pain, and they respond positively to appropriate analgesic and anesthetic techniques.

Pet animals that are in pain appear depressed and cranky. They can sometimes be more aggressive if handled in such a way that exacerbates the pain. Pain shuts down appetite. It stimulates the sympathetic nervous system, resulting in an increased heart rate, increased blood pressure, and sweating. Because that is where their sweat glands are located, dogs and cats sweat only on their paws. Horses sweat over their whole bodies. Humans evidence similar behavioral and physiological responses to pain. Despite minor species differences in expression, such as the precise location of sweat glands, there is little difference in how species respond to pain.

We can clearly see that animals suffer. Descartes should be relegated to the history books and philosophical discussions where he belongs. Animals are not pain-immune automatons. If it looks like pain, behaves like pain, and responds to treatment like pain, then it probably is pain.

The medical-behavioral relationship at times works the other way around, in that emotional and behavioral issues can lead to illness. Factors such as anxiety or stress can actually cause physical disease. Such conditions may be properly labeled psychosomatic. The term speaks to the relation between the psyche, the mind, and the somatic, meaning "of the body."

The power of the mind, in particular one's overall mental attitude, can have positive or negative effects on one's well-being. Whether at the beginning of an ordinary day, or in the midst of chaos, or during an illness, if you believe good will happen, you will notice more good than bad. In terms of positive thinking, if you believe it, it will be more likely to happen. Conversely, negative thoughts and attitudes lead to a lack of resilience and worse outcomes. Specific conditions in humans that are thought to include a psychosomatic component include psoriasis, eczema, stomach ulcers, high blood pressure, and heart disease.

Can animals suffer from psychosomatic conditions, too? Do they have psyches that can be sufficiently distraught as to produce somatic illnesses?

The answer is a resounding yes. As with humans, certain skin conditions in animals appear to be worsened by stress. As we have seen, acral lick dermatitis, while appearing to be a skin disease, is actually an anxiety-driven compulsive disorder with fundamental origins in the brain. That is to say, it is psychosomatic. Other psychosomatic skin conditions include skin picking in monkeys and feather picking in birds.

Anxiety and stress are the driving force behind these behaviors. A condition called neurotic excoriation is a skin scratching disorder reported in humans. I once gave advice to the staff at the Mayo Clinic on how to treat a human sufferer of this condition, who had been refractory to all the treatments the doctors at the clinic had tried.

The patient, a vice president of a major US corporation, neurotically scratched one side of his face, which as a result became deeply ulcerated. He had actually damaged the nostril on that side. Despite the Mayo Clinic's best efforts, his obsessive scratching had been going on for two years.

Within three days of Mayo implementing our recommendation to treat him with an opioid antagonist, naltrexone, the man improved greatly. After a month of treatment, his face had healed. I warrant the same treatment would work for skin-picking monkeys, but I haven't tried that so far. After all, humans are primates, too—the naked ape.

The effects of stress and anxiety on appetite can be quite clear. Dogs eat less and thrive less well when stressed. In extreme cases, illness can follow.

Stress and anxiety can play the same kind of tricks on people, too. Worried humans tend to lose weight and in extreme cases can fail to thrive. Anorexia nervosa is an anxiety-driven compulsive disorder. Sufferers are extremely concerned about their appearance and keep thinking they are overweight.

Stress can affect cardiovascular function in people and animals. Stress causes the release of stress hormones—the very same stress hormones, in fact—causing elevation in blood pressure and heart rate. The long-term effect of these elevations is not good for heart health, particularly in individuals who have heart disease. People do vary a bit from dogs in this respect, however, as dogs tend not to get atherosclerosis. Rather, canines tend to be affected by either valvular heart disease or cardiomyopathy, a disease of the heart muscle.

Robert Sapolsky, a neuroscientist at Stanford University who has studied baboons, found that in-charge, dominant baboons had less stress than those of lower rank. Upon postmortem examination, the subordinate animals often displayed gastric ulcers and enlarged adrenal glands, clear evidence of a psychosomatic manifestation of stress. Years later, through a twist of fate, all the dominant males were wiped out by tuberculosis as a result of scavenging from human garbage dumps. The upshot: a much more peaceful regime, more friendly associations, and, of course, less stress. Let that be a lesson to us all.

So-called irritable bowel syndrome is a gastrointestinal disease of dogs that appears to be psychosomatic. Afflicted animals sometimes suffer explosive diarrhea in stressful situations. There may be other factors involved in this condition, but stress certainly seems to precipitate it, at least in some dogs. I'm sure you know what I'm going to say next. Yes, people get irritable bowel syndrome, which is also exacerbated by anxiety and stress.

Finally, there is type 2 diabetes. A large body of animal studies supports the idea that stress reliably causes a spike of glucose levels in the blood, or hyperglycemia, and aggravates the risk of type 2 diabetes. Hyperglycemia has detrimental effects on nerves. The condition is best controlled by the combination of diet, exercise, and stress reduction. We can be talking about either humans or animals, and the conversation about cause and treatment would be much the same.

Depression of the immune system from stress can lead to a plethora of physical disorders. If a person or animal is under constant stress, hormones such as cortisol will depress the immune function. A depressed immune system in turn leads to increased susceptibility to infections and impaired ability to regulate improper cell division. A healthy immune system, on the other hand, works to mop up aberrant cells that might otherwise multiply, leading to cancer. Cancer is one of the leading causes of human death. It is also the leading medical cause of death in dogs.

Thus psychosomatic conditions exist in both human and non-human animals. Many result from stress and anxiety. Severe ongoing anxiety can affect just about every organ system in the human body. Worry will shorten life span. Clearly, it is important to address the toxic effect of too much stress. Engaging in a healthy lifestyle helps, as does addressing any developing issues before it leads to serious detrimental effects.

Ask a smart aleck about anxiety, and he'll tell you not to worry about it. It is true that stress oftentimes triggers more stress. It can be a self-perpetuating loop. It's important to break the chain for both you and your pet. I once met a medical student at a health spa who was expounding about all the healthy things he was doing to stay in good physical shape. He exercised, did not smoke or drink, and ate only healthy food. But then he added that, unfortunately, he could not stop worrying about maintaining his health (and other things) and he knew that worry, or stress, was the biggest killer of all. Now there's a catch-22 situation if ever I heard of one.

CHAPTER FOURTEEN

The Narcoleptic Horse

Night Terrors and Other Sleep Problems

To sleep, perchance to dream . . .
—WILLIAM SHAKESPEARE

Working with an animal shelter, a caring, well-meaning family accepted a seven-month-old German shorthair pointer puppy named Jake for foster care. The new owners were duly warned that the dog had extreme sleepwalking episodes at least ten times per night, behavior that was probably the reason for his surrender. Sleepwalking didn't sound like that much of a problem to Jake's new interim family, but they soon learned otherwise.

Jake's nighttime behavior also seemed to have no particular trigger. The dog gave little or no warning signs before having one of his "night terror" attacks. Some nights would go by without any, and on other nights they were alarmingly frequent. As each attack started, Jake would paddle furiously whine, cry, snarl, and bark. Spookily enough, at times during these episodes his eyes would be wide open, but he was clearly in some kind of somnambulant state and not fully conscious.

We've seen this kind of nighttime behavior before, in Comet, the Golden retriever brought to me for his inexplicable rages. As with Comet, Jake's foster family also made a video of their dog's strange behavior. But that footage wasn't only utilized to enlighten an animal behaviorist such as myself. Instead, it turned up on YouTube (http://bit .ly/1JOIWXC).

The video got a lot of hits, and comments flooded in. Most of them treated the footage as some sort of amusing spectacle. "He is soooo cute!" read one. "Funny how he jumps up, runs around and plops back down on the pillow like nothing happened." Finally, well down in the comment thread, came a dose of common sense: "This shouldn't be an entertainment for you guys. RBD or REM Behavior Disorder is a serious condition for animals and human. This is pretty severe."

His foster family had found it increasingly difficult to keep Jake for fear that he would hurt a person or another animal. The dog was completely unadoptable at this point. The local vet suspected canine distemper, which causes seizures and tried treating Jake with potassium bromide, an anticonvulsant and sedative. The measure had little success. Diagnosis of such problems in dogs is very difficult. Certainly partial seizures can cause such signs, but so can other sleep disorders. Often the final diagnosis is made retrospectively by positive response to a particular treatment.

Various conditions could have led to Jake's bizarre behavior. One of them is indeed a condition the Internet commentator suggested, RBD (rapid eye movement behavior disorder). After my initial contact with the dog's foster family, I suggested treatment with a Valium-type drug, clonazepam, which is often used for treatment of RBD but also has anticonvulsant properties. A double whammy or twofer deal, so to speak. Unfortunately, I was not able to monitor Jake's future progress because I was not the vet on the spot. I trust that he would have fared well as long as appropriate doses of this medication were employed.

The responses to Jake's video measure the gulf in common knowledge about pet behavior. No one can deny that animals are often comical. David Letterman's "Stupid Pet Tricks" was hugely popular, even though some of the animals featured were actually psychologically disturbed. Observing pets' oddities is one of the great enjoyments of life. At the same time, the ignorance of classic signs of a mental disorder and regarding viewing it as entertainment is troubling. It indicates the distance we have to go informing the public about something so ordinary, so unremarkable, as an interrupted nap.

Nighttime presents many faces, and not all of them are healing. As the all-too-true saying goes, "Four a.m. knows all my secrets." Something about the night brings out our demons. It's the most natural thing in the world, and in humans probably has its roots in our prehistoric ancestors spooking at things that went bump in the night, a survival trait that was good for the perpetuation of the species. In the existential algorithm, darkness so often equals death. It's no wonder that we worry more at night.

In pets, many behavioral issues turn up in the wee hours, some of which have a medical basis, some of which do not. "Nocturnal anxiety" is the catchall term for this condition, which affects pets more commonly than one might think. In my experience with dogs and cats, there are four main causes of nocturnal anxiety: Alzheimer's disease, PTSD, noise phobia, and underlying medical conditions that cause pain or discomfort. I have already discussed the first three conditions in earlier chapters.

The so-called Sundowner Syndrome is one of the first things to cross my mind when an older animal becomes distressed as night falls and darkness arrives. Canine and feline cognitive dysfunction often features changes in sleeping patterns and displays of nighttime anxiety. These are often heralded by the Sundowner Syndrome, which can include disorientation, such as becoming lost in familiar environments;

altered interactions with family members; and barking aimed at nothing in particular.

By contrast, if the distress begins immediately after some traumatic event, PTSD is high on the list of potential causes. For noise phobia to be involved, the sound the pet is fearful of may be more clearly audible at night because of the lack of masking noise of daytime clamor. One dog I treated had nocturnal panic attacks, triggered by the screech of the metal snowplows across the road surface, setting the poor pooch's nerves on edge. Another dog became fearful of the noise of the boiler turning on at night. More recently, I've seen a dog whose trigger was the sound of a pack of agitated coyotes howling nearby at night.

It is fairly easy to ascertain a diagnosis for such conditions in dogs or cats. What is more difficult to diagnose is the effect of some painful disease that causes an animal distress at night. We have to gradually exclude other causes in order to make a diagnosis.

I discovered the ins and outs of this situation by accident many years ago while treating an aged Afghan hound for nocturnal anxiety. Our internal medicine vets at Tufts had gone over this dog with a proverbial fine-tooth comb. They could discover nothing to explain his sudden change of behavior.

The dog was turned over to me, in the hope that I might come up with some therapeutic cocktail that would make him more restful and less anxious at night. This I happily did, using a combination of a mild antianxiety drug, buspirone, combined with a morphine-like analgesic, butorphanol. The results of the treatment were good. Within days the dog and his owners were once again sleeping through the night.

A few months later, the dog had a spontaneous fracture of one of his legs. When he was admitted back to our hospital, his leg was X-rayed and then, because of suspicious findings, he was sent for a nuclear bone scan. This test, which also goes by the name scintigraphy, finds anomalies in bone. It's sophisticated, expensive, and obviously not part of a routine examination.

Unfortunately, the scan turned up a serious problem. The hound was found to have multiple myeloma, a terminal condition. Like some malicious Pac-Man, the cancer was eating away at his bones.

From cases like these, I learned over the years that nocturnal anxiety can be an early sign of deeper medical problems, an insight that has helped me diagnose and treat dogs in similar predicaments. As I slowly accrued a number of cases of medically induced nocturnal anxiety I became more confident about the diagnosis. In the majority of cases, the medical cause of this condition turned out to be some form of cancer.

One older dog with this condition clearly did not have Alzheimer's or PTSD. He was not sound sensitive. By eliminating these common causes of behavioral changes in the nighttime, I was able to zero in on medical issues. The dog turned out to have a retrobulbar tumor, a malignant growth behind his eye. At the time of the dog's initial clinic appointment, I did not know what the medical basis of his problem was, but I felt confident enough to inform the owner that I was sure his pet had some medical issue causing his nighttime anxiety. That problem, I said, would become apparent in time. I treated the dog's symptoms and waited.

Sure enough, some months later, his owners reported that his eye had started to bulge out of its socket. They contacted an eye specialist, who removed the retrobulbar tumor. The dog's nighttime anxiety abated. The dog understood he was unwell long before we figured out what was wrong with him.

If nighttime anxiety at times occurs in a dog with a painful medical condition, it stands to reason that the same must also occur in people. I did not know this for certain at the time I was treating dogs, but could have predicted it by applying the principles of One Medicine. After years of diagnosing medically induced nocturnal anxiety in numerous canines, I encountered a client who clued me in.

"They give a handout to parents of children with cancer who are being treated at the Dana-Farber Hospital in Boston to warn them about nocturnal anxiety," the client told me. "That's when things seem to get so much worse." Apparently, children with cancer are more likely to suffer distress at night than at any other time in the twenty-four-hour cycle.

The reason for this, I assume, is the same as in dogs with painful medical conditions. At night there is little else going on. There are few lights or sounds for distraction. As is said in The Who's famous pop song "Pinball Wizard":

> Ain't got no distractions,
> Can't hear no buzzers and bells.

Family members might be all asleep, the nursing staff is reduced to bare-bones level, so human interactions become more limited when compared to the daylight hours. The mind is free to focus on internal processes.

In pets as in humans, this situation causes insomnia, restlessness, and distress. Such conditions generally disturb the night, for the sufferer as well as the other members of the household. The ultimate solution is to surgically remove the offending tumor, if possible, or treat it with radiation or chemotherapy. In the meantime, as with the Afghan hound, anxiolytics and analgesics are helpful in treating symptoms.

Sleep apnea can occur in both people and dogs. Most affected dogs are short-nosed dogs such as English bulldogs, Pekes, boxers, and Boston terriers, who suffer from what is known as the brachycephalic airway syndrome (BAS), an unwelcome spin-off of selective breeding. Physical aspects of BAS include an elongated soft palate, an underdeveloped trachea, everted laryngeal saccules, and stenotic nares, which is the medical term for narrowed nostrils.

Though not all brachycephalic dogs have the full gamut of physical issues, even two or three of the symptoms can be compromising. Veterinarians with a special interest in bulldogs must learn to be proficient at corrective soft palate surgery, a procedure similar to the radical throat procedure performed in people who snore and have obstructive sleep apnea.

The brachycephalic syndrome can be problematic during the day also, especially in hot humid weather or when dogs with the condition are exercised hard. Some even collapse and have to be revived. Nighttime brings its own problems, as these dogs wheeze and struggle for breath. "Snore and you sleep alone," is the old adage, and for this reason many short-nosed, snorty, and snuffly dogs wind up exiled from place of pride in the bedroom.

Signs of sleep apnea in dogs are the same as those in people. They include loud and chronic snoring, interrupted breathing with long pauses between breaths, choking or gasping during sleep, frequent waking during the night, daytime fatigue, irritability, aggression, and depressed mood. As with people, being overweight increases the severity of the condition.

So if you have a cranky old, overweight English bulldog—or a cranky old overweight husband—airway problems may be partly to blame. For husbands, nasal surgery may help to correct the problem, and for dogs such a procedure can be performed to correct genetic malformation of the nose cartilage.

A less invasive alternative to surgery is continuous positive airway pressure, or CPAP, via a nasal or full face mask. Currently only people will tolerate these appliances, and no genius inventor has come up with a viable CPAP rig for dogs. A last resort for dogs might be throat surgery. Some people become so worn down by their dog's sleep apnea that an invasive surgical approach finally becomes the lesser of two evils.

The science fiction author Philip K. Dick posed a novel question as the title of one his books, *Do Androids Dream of Electric Sheep?* Like anxiety or post-traumatic stress, dreaming is one more facility that some people believe only sophisticated human brains have. A simple glance at a dog's brain waves during sleep would disabuse such advocates of human exceptionalism.

Dogs do dream. Like our dreams, theirs come during a phase called rapid eye movement sleep. During this REM phase, the large antigravity muscles of the body are mostly paralyzed, but muscles that control fine motor movement remain active. In dogs, these include the muscles that control the movement of the eyeballs, eyelids, muzzle, ears, larynx, whiskers, and paws. Just as with people, a dog's eyeballs move rapidly to and fro. Because of the muscle paralysis and the resulting relaxation, the REM phase is sometimes called the sleep of the body. But during REM, brain-wave activity remains decidedly active. Neural activity is similar to that seen in animals that are awake. It is in this phase of sleep that dreaming occurs.

Once in a while, for both people and dogs, the large muscles are not paralyzed in sleep. Though soundly asleep, the animal can physically act during his dreams. Like most dogs with REM behavior disorder, people with this affliction may exhibit mindless aggression. Dream-enacting behaviors in people include talking, yelling, punching, kicking, sitting, jumping from bed, arm flailing, and grabbing. Similarities between the human and canine condition abound.

Waking behavior during sleep is one thing. Sleep behavior during wakefulness is the flip side of the coin, and goes by the name of narcolepsy. Narcolepsy causes animals or people to fall asleep suddenly during the day, sometimes in the middle of an activity or meal. The sufferer feels excessive daytime sleepiness, punctuated by sudden bouts of REM sleep.

The narcoleptic might also collapse into a conscious, paralytic state called cataplexy. Excitement caused by, for example, anticipation of food or other upcoming events is often the trigger for these so-called sleeping attacks.

Years ago I was consulted about a horse that was literally falling asleep on his feet. The quarter horse would buckle at the knees and almost collapse when his girth was cinched prior to going out on a ride. Because of this stumbling and falling behavior, the owner was quite rightfully concerned about possible danger to her daughter, who often rode the horse. She was considering putting the animal down.

Tufts veterinary school was this owner's last resort and the horse's last chance. Of course, more than one condition can cause a horse to collapse, so we brought the animal into our equine wards for a few days to have various specialists take a look at him and put him through a complete gamut of tests.

Internal medicine specialists found no diseases that would cause him to collapse. Neurologists declared there were no signs of brain conditions. Cardiologists checked the horse's heart and employed a "Holter monitor" for continuous ECG monitoring, which ruled out intermittent cardiac rhythm disturbances. Since these investigations were all negative, the case was passed off to me, aka "End-of-the-Line Dodman."

I suspected that the horse had narcolepsy, and I treated him with a tricyclic antidepressant, amitriptyline. We waited and watched. On this medication, the horse had no further attacks, not even when the girth was cinched. The horse went home on a daily dose of the medication. For as long as I followed up with this owner, the horse had no further attacks.

Studies of a pack of narcoleptic dogs at Stanford University in the 1970s led to discovery of a genetic glitch associated with the condition. A research team headed up by Emmanuel Mignot, a French pharmacologist and director of the Stanford Center for Sleep Sciences and Medicine, found a mutation of a gene responsible for proper formation of a brain chemical that regulates wakefulness and appetite.

Mignot and his wife, Servane Briand, adopted one of the dogs in the study, Bear, a male Belgian schipperke. The dog was brought to Mignot by a breeder, who told him the animal "fell down every time he got excited." Bearichon, as he was affectionately called, lived to a ripe old age, a further indication that narcolepsy isn't a life-threatening affliction—unless a person falls asleep in the middle of the street or while driving.

Enhanced emotions trigger narcoleptic episodes in dogs. Bear fell asleep whenever he got particularly animated. While he was eating he would collapse because he'd suddenly get excited about his meal. In other studies, laboratory dogs fell asleep when shown an escape route to the exterior. Free, free at last! Only just let me take a little nappy first. People may have narcoleptic attacks triggered by strong emotions, too. They often occur when people are laughing or crying.

Treatment of narcolepsy is the same for animals and people. One of the mainstays is a stimulant drug called Provigil, named from the Greek *pro*, for "vigil," being watchful. Another treatment involves serotonin and norepinephrine-enhancing tricyclic antidepressants, such as the one we used in the narcoleptic horse case mentioned above.

Shakespeare calls sleep "nature's soft nurse," yet it proves problematic in such conditions as night terrors and narcolepsy, the two ends of the sleep disturbance spectrum.

CHAPTER FIFTEEN

The Listless Pet

The Thyroid and Anxiety, Aggression, and Mood

The cure of the part should not be attempted without
the cure of the whole.

—PLATO

A few years back, I listened to a woman speak on NPR about some issues she was having with her memory and mood. She kept losing her car keys, she said, which was not normal for her, and she was often tired. In the most general terms, she simply did not feel right. She knew something was wrong and, for some reason, suspected her thyroid was out of whack. Her doctor's test for thyroid dysfunction was negative, and he reported that all values were "within normal limits."

The woman did not quit in her quest to find the source of her difficulties. She asked to be referred to an endocrinologist and her doctor obliged. On rerunning the tests, the endocrinologist diagnosed subclinical hypothyroidism, a borderline state of thyroid dysfunction in which thyroid hormones are within the normal range but suboptimal for normal functioning.

The specialist treated the woman with thyroid hormone replacement therapy, prescribing levothyroxine. Her condition immediately improved. No more funky mood swings, no more memory loss, no more tiredness and irritability. The transformation was profound.

With impressive communication skills and determination, this woman managed to help the medical profession help her. Unlike a doctor interviewing such a patient, veterinarians cannot question animals about what ails them. "Do you remember where you buried your bone?" isn't a viable clinical approach to check for memory loss in dogs, nor is "Are you worried all the time?" when seeking signs of anxiety in cats.

Instead, we veterinarians must figure out our patients' health from observing them. And we can miss the nuances in the presentations of our patients. A condition such as borderline hypothyroidism, for example, is difficult to nail down in people, much less pets. People with depression, anxiety, and other psychiatric problems often have abnormal or low blood levels of thyroid hormones. And treating the thyroid problem can lead to improvements in mood, memory, and cognition.

Hormones secreted by the thyroid gland act as the central nervous system's traffic cops, functioning as neuromodulators and neuroregulators. Thyroid hormones also influence other neurotransmitters like serotonin, norepinephrine, and dopamine, helping to control mood and behavior. Low levels of thyroid hormones lead to decreased serotonin, decreased norepinephrine, and reduced dopamine levels in the brain.

None of these effects is good. Low levels of serotonin are associated with depression, sleeplessness, and anxiety. Low levels of the catecholamines, norepinephrine, and dopamine, both "go" neurochemicals, set a person up for tiredness and depression. Dopamine is involved in the brain's natural reward system. With too little of it around, the impetus to engage in pleasurable activities is muted.

Most medical professionals quizzed about the mental status of someone with hypothyroidism would invoke words like "depressed, lacking energy, lethargic." That is true of full-blown hypothyroidism, often referred to as "frank" hypothyroidism. Signs of the borderline condition, on the other hand, are more subtle. They include memory impairment, anxiety, as well as behaviors that derive from these impairments. Correction of borderline or low thyroid status often completely reverses this literally depressing situation.

In people and pets, borderline hypothyroidism can also lead to paradoxical effects of agitation, paranoia, and aggressiveness, which can be confusing for physicians and vets, who can easily miss the right diagnosis. Is Fido just getting old, or is there something physically wrong that is causing her listlessness? Making the leap from behavioral condition to a medical issue underlying it can be tricky, but there is often a direct link. Other common medical problems in pets, such as partial seizures and Alzheimer's disease, have behavioral consequences, and animals and people who are ill may retreat into themselves, act depressed, or become extra needy. The psychosomatic effects of excessive anxiety in pets can cause very real physical problems, as in the case of irritable bowel syndrome and acral lick dermatitis.

Veterinary students have an acronym to help them to remember the various medical conditions they may encounter. It is referred to as DAMN-IT:

- *D = Degenerative or Developmental*
- *A = Anomalous or Autoimmune*
- *M = Metabolic, Mechanical, or Mental*
- *N = Nutritional or Neoplastic*
- *I = Inflammatory, Infectious, Ischemic, Immune-mediated, Inherited, Iatrogenic, or Idiopathic*
- *T = Traumatic or Toxic*

Any of these conditions may affect behavior. A prime example is a metabolic one, hypothyroidism, the most common genetic disease of purebred dogs. It is the number one health concern in six of the seven AKC-recognized breed groups. Since hypothyroidism is so prevalent, then veterinarians and behaviorists should reasonably expect to encounter not only frank hypothyroidism on a regular basis, which we do, but also borderline hypothyroidism.

The first time I encountered hypothyroidism contributing to a psychological condition in dogs was many years ago. A family of Afghan hounds who were regularly shown by their owners as fine examples of the breed had anxiety problems. From a young age, these dogs went through mock show exercises to prepare them for real events, but at some point between nine and eighteen months of age, some of the Afghans began showing extreme anxiety when approached or examined by the show judge. Before a certain age the dogs were perfectly fine in the mock show situation. Then, as they crossed an age threshold, it was as though a light switch had been thrown. They became anxious.

Blood tests showed that the Afghans had low-to-borderline thyroid hormone levels. When these dogs were treated with a supplementary thyroid hormone, levothyroxine, their anxiety dissipated. All was well on the exhibition circuit.

Hypothyroidism gradually became recognized by other experts as causing anxiety and other anxiety-related behaviors. It's a great feeling, when you can sense the scientific community closing in on a problem. Soon word on the street among dog trainers and veterinarians was that borderline-to-low thyroid hormone levels were associated with anxiety and aggression.

With all this in mind, I started to look more closely into behavior cases that were presented to our clinic at Tufts. Dogs who were anxious or aggressive, especially when those behaviors were associated with subtle physical signs of hypothyroidism, now prompted me to perform a thorough thyroid hormone analysis.

Clinical signs that tipped me off to a possible diagnosis of borderline hypothyroidism included excessive shedding, premature graying of the muzzle, a sad facial expression, drooping lower eyelids, and slow hair regrowth of previously shaved areas. Weight gain was another indicator. Dogs with borderline thyroid felt cold all the time and sought out sunlight, heaters, or other warm areas. They exhibited low body temperature, slow resting heart rate, and skin thickening, especially in areas of skin-to-skin contact such as armpits. Skin thickening also causes drooping in the facial region, contributing to a "sad" expression.

For patients with these symptoms, I ordered a blood test and, if results indicated that thyroid levels were borderline low range, I treated the dogs with hormone replacement. Any ranking that was in the bottom tenth percentile of the normal range was cause for concern. Many dogs improved dramatically in as little as five days. Other times, improvements took as long as six weeks.

One case that I clearly remember involved a German shepherd who displayed aggressive behavior to puppies, other adult dogs, and to strangers, but showed absolutely no physical signs of hypothyroidism. The owner had been to vets and trainers everywhere, done everything she knew she should, and spent all kinds of money in vain attempts to resolve her dog's behavior problems.

I decided to check the dog's hormone blood levels in case the previous examinations had missed something. Out of the 177 AKC-recognized breeds, German shepherds rank number seven on the hit parade for hypothyroidism. I did not want to leave any stone unturned with this dog. The owner was desperate.

The dog's thyroid levels came back in somewhat low, but still in what formerly was deemed an acceptable range. I was in the same position as the woman I had heard about on NPR, whose thyroid hormone levels were officially medically normal. I initiated treatment with thyroid hormone replacement anyway.

The dog responded brilliantly and became a changed animal. Five days after his thyroid hormone treatment began, the owner took the dog on a walk. The shepherd played with a puppy, engaged in friendly tug-of-war with another dog, and allowed himself to be approached by strangers. The owner emailed to tell me that she cried tears of joy after that walk. After two years of input from multiple trainers and veterinarians, this was the first concrete progress she had seen. And the dog's improvement lasted.

Another dog I treated was violently aggressive to all other dogs. As with the German shepherd, I checked his thyroid levels, found them to be lowish and treated him with thyroid replacement therapy. That was all this dog needed. His aggression toward other dogs completely disappeared, simply by correcting his thyroid status.

I wrote up a case report, documenting three aggressive dogs responding well to thyroid hormone replacement, which was featured as the Behavior Case of the Month in the *Journal of the American Veterinary Medical Association*. I hoped this might lead some of my colleagues to pursue a similar diagnosis with their behavior cases, but there were skeptics. Some veterinarians simply doubted the existence of borderline hypothyroidism as a cause of canine behavioral problems and aggression. Others saw the issue in binary terms, either/or, black or white—no gray scale allowed, if you please. Such binomial thinkers recognized only two possible findings, normal and low thyroid status. There would be no borderline interpretations for them.

It is easy to recognize the robustly normal thyroid state in a dog with no physical problems, no behavioral problems, and thyroid hormone levels in the mid- to upper end of the normal range. Also obvious is frank hypothyroidism, since blood work will show hormone levels to be well below the lower limit of normal. These dogs often display other signs, such as being overweight, lethargic, apathetic, or having bilaterally symmetrical hair loss. Their condition can be diagnosed from the top of a double-decker bus with a telescope turned the wrong way around.

I don't believe that there is only door number one and door number two, normal or abnormal. I believe there is something in-between, a door number one and a half, if you will. It occurs in people, as in the lady on NPR. One Medicine predicts it occurs in pets. I call this condition borderline or subclinical hypothyroidism, and I treat it.

In an attempt to address the critics of canine borderline hypothyroid, I set about conducting a double-blind, placebo-controlled study of the effects of treating borderline hypothyroidism in aggressive dogs. To enroll in the study, dogs had to show mild to moderate owner-directed aggression, have one or two clinical signs of hypothyroidism, and have a thyroid hormone levels at the lower end of the normal range. They were treated either with L-thyroxine or identically shaped and colored placebo tablets. The results of this study showed that the frequency of aggressive events was significantly lower in the thyroid hormone–treated group versus the placebo group after six weeks of therapy.

While this sounds fairly conclusive, I must admit there's a caveat. None of the dogs in the study showed the same dramatic responses that I saw from time to time in the clinic. While the jury might still be out on borderline hypothyroidism, I firmly believe it exists. Clinical evidence, at least, is mounting that dogs can have the borderline syndrome of subclinical hypothyroidism. Because affected dogs show increased anxiety, aggression, and skittishness, and aggression is a leading cause of dogs being put to sleep, it is vital that these avenues of research be pursued.

With hyperthyroidism—elevated thyroid levels, as to the opposite problem of hypothyroidism—there are few controversial issues at all. People suffer from hyperthyroidism, and cats frequently become hyperthyroid when they are elderly. Occasionally, dogs with thyroid tumors become hyperthyroid.

Hyperthyroidism occurs when the thyroid gland is overactive and produces too much thyroid hormone. I always recall the Mark Twain line: "Too much of anything is bad, but too much good whiskey is barely enough." Well, too much thyroid hormone in the bloodstreams of both people and pets can be disastrous.

Graves' disease is the most common cause of hyperthyroidism in people. It is caused by the aberrant production of antibodies that mimic a regulatory hormone secreted by the pituitary gland.

In cats, hyperthyroidism is one of the more common conditions of older felines. Ingestion of certain soy-based proteins found in some commercially available canned cat food is a possible cause of hyperthyroidism in cats. Another theory is that it results from the ingestion of flame retardants that are now ubiquitous in home furnishings, including rugs, upholstery, and curtains. These chemicals eventually wear off the furnishings and become incorporated into house dust, which then gets in the cat's fur, and the cat licks them off when grooming.

Signs of hyperthyroidism, both physical and behavioral, are common across the species, including high blood pressure and rapid heart rate. Behavioral signs include irritability and aggression, excessive reaction to stimuli and increased appetite, paradoxically accompanied by weight loss. Treatment by means of drugs, surgery, or radiation therapy is effective at reversing these signs. I have treated several hyperthyroid cats. One that is most memorable to me is Baby, a cat who lived with me for a period of time toward the end of his life. Baby's owner had died. I inherited the cat, getting nominated for the job as one of a few people who could have cared for him because of his aggressive behavior.

In his previous home, Baby was hyperkinetic, hypervocal, highly irritable, and extremely aggressive. He also had an increased appetite, was skinny as a rail, and had a disheveled hair coat. I diagnosed him as having hyperthyroidism, and began treatment with a medicine called Tapazole. This twice-a-day drug stops the thyroid gland from making too much thyroid hormone. Even on this treatment, Baby remained somewhat feisty. But many of his other behavioral signs improved.

I recall that getting the treatment into him proved a bit of a chore. My wife, Linda, solved the problem in an ingenious manner. She simply grabbed him, placed him on the corner of my daughter's bed, and pilled him. Then—and this was the most important step in the process—she fed him. It was not long before he jumped on the bed and waited eagerly for his pill, because it signaled the arrival of food.

I once diagnosed hyperthyroidism in a dog by accident. I was leaving the gym one day when a friend spotted me across a parking lot.

"Hey, Nick, my dog has started to act aggressive to other dogs," he called out. "I have no idea what to do. Do you have any quick advice?"

As I was about to step into my car at the time, I shouted back the first thing that came to my mind. "Take him to your vet and get his thyroid checked. Let me know what you find."

A few months later, I met the man's wife at the same gym. She ran up to me, threw her arms around me and gave me a big hug. "Thank you so very much," she said. "You hit the nail right on the head. Our dog's problem was a result of a thyroid issue. If you hadn't suggested we check his thyroid gland we might never have found out."

She went on to explain that the dog's thyroid level had tested out extremely high. He proved to have a thyroid tumor, which was operated upon. Recovered, the dog now was behaving like his old self. Since I was expecting hypothyroidism to be the diagnosis, but had not imagined a tumor could be the cause, this came as quite a shock. I simply said, "Happy I could help."

The good news is that hyperthyroidism symptoms across the species are pretty much the same. Dogs and cats do not need to have language in order for vets to see what is happening. And the condition improves with the same treatments in both human and nonhuman animals alike.

CHAPTER SIXTEEN

Blue Dogs and Cats

Depression and Mood Medicines

An animal's eyes have the power to speak a great language.

—MARTIN BUBER

Many people are surprised when they ask me whether pet animals can experience depression, and I respond, "Yes, of course." Depression certainly can and does occur in pets. A complex mental condition, depression is not a psychological state exclusive to humans. Naturally, because we are unable to communicate verbally with animals, we do not know as much about how depression manifests in animals as we do about how it affects people. For the same reason, the classification of depression in animals is somewhat imprecise. We can put a pet on a couch, so to speak, but our notes on the session won't include any verbal replies. We can, however, verify that the same molecules of emotion are coursing through a pet as through a person.

One of the first cases of animal depression I ever saw involved a cat named Maxwell. A few weeks before visiting me, Max's beloved housemate, a companion cat with whom he was closely bonded, had died.

Max went into an extreme behavioral funk. He refused to eat, became morose, lost interest in all normal activities, and basically slept the day away. As a result, Maxwell lost a great deal of weight. His owner initially brought him into our veterinary hospital to see one of the doctors in the Internal Medicine Department, who diagnosed liver failure, an unfortunate and extremely serious consequence of rapid weight loss in felines. The Tufts vets brought Max into our intensive care unit, where he was hydrated with intravenous fluids and received state-of-the-art treatment for liver failure. Although such a condition usually carries a grave prognosis, we managed to pull Max through.

Fortunately for Max and his own er, the clinician was savvy enough to realize that the root cause of the problem was depression-related anorexia and referred them to me for advice on how to manage Max's mood, as he was still depressed. In order to help lure him back into the fullness of life, the owner needed to give him even more caring attention, to generally enrich his home environment with interactive toys and games, and possibly adopt a new kitten to help make up for the loss of his dear friend. The mainstay of my treatment, however, was Elavil, an antidepressant commonly used for people. After a couple weeks, Maxwell was back on his feet, eating and living a normal life, although the owner did not take my advice to get a new kitten.

To me, Maxwell is a poster child—or poster cat—for how psychological issues can affect physical health and wellness. In psychiatry, this feline's experience would be termed a major depressive incident. Some stern stoics believe that the blues can simply be shrugged off with strongmindedness. "It's a question of character, damn it!" But Maxwell, the cat with the blues, shows that a creature can come close to dying from the physical repercussions of depression.

One of the several dogs I have seen with depression also developed the condition following the loss of a beloved housemate. Like Max, Oliver, a neutered mixed-breed pit bull, lost his appetite, became uninterested in toys and games, and generally lost interest in life. He moped around the house and slept for inordinate amounts of time.

I treated the dog in a similar way to the cat and recommended what we might call in the UK "jollying up" routines. I told the owners to act happy around Oliver, to jolly up Ollie, so to speak. High spirits can be as contagious as low ones. Plus I advised that they play with him a lot more, devise games, and give him attention, delicious food, walks, and other outdoor activities. I also advised the family to consider adding a new pet to their household.

In fairly short order, the owners did all of that. Though Ollie's depression had prevailed for several weeks, upon arrival of a new puppy he began to turn around quickly. Soon Oliver was completely back to his old self. The new puppy, in this case, was just what the doctor ordered.

Another canine case had special, heartbreaking circumstances. The owners had tried to hide the death of one of their dogs from the surviving pet, who had been closely bonded with his housemate. The owners had had to euthanize the ill dog to prevent his increased suffering from a terminal disease. The wife took the healthy dog away from the house while the owner's spouse, who had taken their sick dog to the vet for the sad procedure, brought the body back home and buried it in a deep grave in the backyard.

When the euthanized dog's canine companion returned home, he ran around the house looking for his old friend, dashing from room to room. Eventually, he went outside and tore around the perimeter of the house in a state of panic. When he finally discovered the fresh grave, he lay on top of the dirt mound and would not budge for several days.

Examples of dogs who mourn their owners abound. One of the most famous stories is of a little Skye terrier in nineteenth-century Edinburgh, Scotland. Known forever after as Greyfriars Bobby, the dog faithfully guarded the grave of his owner for the last fourteen years of his life. A book and a movie have been devoted to that remarkable case as well as to that of the Akita, Hachiko, celebrated in Japan as a model of loyalty. Hachiko always saw off his owner at the train station and went back to meet him at the end of the day. When the owner failed to return one afternoon, having suffered a stroke and died while at work, Hachiko met the same train again and again for nine years, waiting for his owner's return.

Clearly, most mourning dogs miss their dearly departed and suffer from a depressed mood. The condition is, for the most part, not life-threatening, and can probably be classed as minor depression, but it can last for quite some time.

Horses also feel depression as a result of bereavement. One of our own horses, Dolly, a chestnut-colored mare, showed signs of such depression. The love of her life was our other horse, Jack, a jet-black, 1,500-pound Cheval Canadian gelding with a flowing mane. Dolly adored Jack and Jack adored Dolly back. It was as if she only had eyes for him and he only had eyes for her. My wife and I had owned these two horses for many years. Jack once had a bout of colic but otherwise was in good shape. As they aged we became progressively more cognizant of the fact that they would not be around forever. Jack began to have arthritic issues and was almost always lame or sore on one foot or another. Corrective shoeing and a regimen of painkillers became the order of the day.

Jack was a proud and handsome beast. Dolly knew it and admired him for what he was. But one terrible day, when Jack was about twenty-five years old, he colicked again, this time much more seriously. My veterinarian wife, Linda, knows a fair bit about horses and recognized immediately that Jack needed an equine specialist vet, whom she summoned. The vet administered painkillers, performed a thorough examination, and took blood for analysis. Toward the end of that day, Linda was quite hopeful that he might turn a corner.

The next morning the phone rang early and we learned the awful news that Jack had cast himself on the barn floor and was thrashing around wildly. Linda hopped in her car and sped to the barn, calling the equine vet on the way. They arrived at the same time. They gave him more painkillers and another examination. The news was not good. Jack's impacted large colon had now shifted horizontally across his abdominal cavity, a very bad sign.

Linda called to tell me that she was going to have Jack put down because he was in so much pain. Even with emergency surgery, his chances of survival were next to none. I tried to get her to bring Jack to Tufts Large Animal Hospital, but she knew I was grasping at straws. Jack was peacefully euthanized at the barn and buried.

For weeks after Jack's death, Dolly moped around and was clearly not herself. She did not interact with the other horses when turned out to pasture but simply stood there, limp, listless, staring blankly at nothing in particular. Her stall was opposite Jack's and so now she had only an empty space to look at. As soon as we saw that depressing situation we had another horse moved into Jack's former stall. Dolly's behavior hardly changed. It was another horse all right, but it wasn't Jack.

A lot of people say to friends who have lost an animal, "Oh, don't worry, you can always get another pet like Pookie." But you can't. Every animal is an individual, so substituting one for another is not something that instantly relieves the pain.

The same holds true for horses. After weeks of this low-ebb state, Dolly did come out of her funk and eventually got back to life as usual. But during her "mourning period," it was plain—and painful—to see her suffering over losing Jack. Here was a clear demonstration of a secondary emotion in an animal: Dolly's primary emotion of sadness led to such secondary emotions as suffering and loneliness. Since such a reality cannot be easily quantified, some people continue to deny that it occurs. But we saw our mare bereft and isolated, a solitary being standing apart in a pasture full of horses, missing the one horse who was her soul mate.

Other forms of depression in animals may be more subtle. Some house cats can become depressed when they lack exercise and the opportunity to engage in psychologically gratifying, species-typical behaviors. Cardinal signs of this type of depression are that the cat sleeps for hours a day, eats a lot, and is overweight. Overeating and weight gain, the antithesis of anorexia and weight loss, can also be signs of depression. It is absolutely not normal for house cats to sleep for 75–80 percent of the time. Nondepressed cats sleep for only around ten hours a day. When the period of sleep climbs higher than that, the behavior becomes abnormal and should be examined as a possible psychological problem. While depression may be too strong a term for this chronic state of affairs, behaviors like overeating and oversleeping are definitely cause for concern.

A better term to invoke here might be dysthymic disorder. In humans, the condition is characterized as "depressive-like symptoms" that are severe enough to interfere with normal functioning and well-being. The correct treatment for what might be called feline dysthymic disorder is somewhat counterintuitive.

I do not advise owners to allow cats outside the house. The outdoor environment is dangerous and life-threatening. Instead, I suggest owners provide their cat a good deal of indoor enrichment and entertainment, as well as encouraging daily exercise. A compatible feline companion cat can also help a morose, inactive, overweight cat by providing opportunities for social interaction and play, both of which can help chase the blues away.

When deprived of opportunities to interact and behave normally, dogs also may suffer from a chronic disorder of depressed or dysthymic mood. The classic example here is, once again, the brood bitch in a puppy mill. The typical subject spends a large percentage of her life crated. She has little opportunity to do anything except eat, drink, and lie down or stand up. Stress, isolation, and lack of social support are the order of the day for these pathetic creatures. In his former existence, when he was a twenty-three-hour-a-day prisoner in his own crate, my dog, Jasper, slept away his days.

Such dogs seem to exhibit what is called learned helplessness, which is representative of human depression. Their barren existence can lead to dysthymia or other stress-related conditions. Many dogs rescued from chain-link prisons arrive in a new home in a completely dysfunctional state. As mentioned previously, they do not bark, play, or wag their tails for many months after they are "sprung," and appear socially anxious and generally fearful, although treatment and retraining, such as my wife and I gave our dog Jasper, can often help them recover.

A tragic variation on the theme of learned helplessness is evident in the story of a gorgeous English gun dog, who was accidentally locked in a pantry while his loving family was preparing to go on a long road trip. It was a bit like the movie *Home Alone*, when in the midst of a confused departure, a family takes off leaving their young son, well, home alone. The difference is that the true story had a much darker conclusion. When the family returned weeks later, they found their dog had starved to death.

The original plan was for the setter to have free entrance into the house and egress via a pet door to the yard, where he would have access to a self-feeder. Yet even in the pantry, the dog was surrounded by food. Game birds had been hung—within reach—to ripen and plastic bags full of food could have been easily ripped open.

When I first heard of this case, I concluded that the dog did not eat the food out of sheer obedience, having been trained not to touch the spoils. But now, years later, I believe the poor dog found himself alone in a hopeless situation. His reaction morphed from panic to helplessness to depression. He went into a state dog trainers call "shutdown." This is seen occasionally during harsh leash training, when canines collapse into atonic immobility after receiving a barrage of inescapable "corrections." Shutting down is really a kind of learned helplessness, a condition that is a dead ringer for—and used as a model of—depression.

This happens when there's no way out. Whatever you do, whichever way you turn, nothing will change the terrible situation you find yourself in. Appetite is lost, motivation is lost, activity is reduced, numbness follows. Although this story had a terrible ending, it serves to show that dogs—and other animals—sequestered at home during long owner absences can become frankly depressed. And in such cases we can objectively prove the anguish that such animals feel, since research studies have turned up increased stress hormones in their blood. Eighty percent of dogs left home alone have increased cortisol levels, and cortisol is one of the stress markers in dogs and people.

The next time your sitter or the kennel owner tells you that your dog merely sleeps while you are away, think again. He may be depressed.

Psychiatrists have long argued over whether depressive personality disorder exists and whether it can set someone up for depression. A gloomy, dejected outlook, worry, and pessimism can predispose people to depression. I can think of animals that I have seen who have such a negative outlook on life, dogs in particular and horses, too. I'm not thinking of Droopy or Eeyore here, but actual animals that I have known. One owner actually brought me his dog because the dog appeared to him to be depressed all the time. There was no known precipitating cause for this dog's gloomy outlook on life. The dog stood like a sawhorse in my consulting room and did—nothing. Facially, he had an unmistakable hangdog look. Nothing I could do at the time—showing him novel dog toys, offering him special treats, petting him—caused him to show any interest. I did my best to advise this owner how to proceed, providing his dog more exercise, environmental enrichment, special food, mano a mano benign training at home, and more social opportunities. The treatment advice barely worked for this dog and the owner did not want to pursue pharmacologic intervention, so we both had to accept the fact that this dog's negative outlook on life was simply a result of his innate personality.

In humans, another personality trait that may predispose people to depression is neuroticism, which is characterized by anxiety, moodiness, worry, envy, and jealousy. Personally, I believe that certain personality traits in animals do predispose them to depressive tendencies, but nailing down what that is has proved an elusive goal. Yet, the canine purebred population represents a virtual genetic laboratory, one that rivals Gregor Mendel's pea garden in potential usefulness to science. If such a predisposition toward depressive illness could be confirmed within certain breeds, then such populations of depression-sensitive dogs could help scientists searching for the genetic underpinnings of depression in people.

Depressed animals that I have seen do indeed seem to be more anxious and prone to worry than their more confident peers. I have not made a specific study of this subject, but my intuition and experience tell me that a dog who is morbidly attached to its owner or to another dog, or one that displays separation anxiety or other anxious conditions, may be more prone to dysthymic disorder or depression. That said, if environmental pressures are extreme enough, most animals, including humans, can be driven to depression.

In 2009, France's former president, Jacques Chirac, and his constant Maltese terrier companion, Sumo, went through an acrimonious breakup. The dog had undergone treatment for depression after Chirac yielded the French presidency to Nicolas Sarkozy, and both pooch and owner were turned out of the Elysée Palace. The media questioned whether canines can indeed suffer from the condition, but it sure seemed as though Sumo wrestled with depression, so to speak. Whether from the abrupt change in surroundings or the lost election, the dog behaved lethargically and lost his appetite. He also became increasingly snappish. Accustomed to roaming free in the expansive gardens of the Elysée, the terrier found himself confined in an apartment on the Quai Voltaire. Luxurious as the new digs were, it was no palace. The formerly mellow Sumo bit the ex-president two times, causing serious enough wounds that Chirac required medical attention. Twice bitten, once shy. Chirac exiled Sumo to a farm owned by friends.

Sumo's symptoms were eminently familiar to me: the lethargy, lack of appetite, the previously unseen edginess and aggression. At Tufts in the 1980s, decades before President Chirac lost his lovely terrier, I was at the Animal Behavior Clinic, and a colleague and I examined a dog with similar symptoms to Sumo's—and who had also bitten its owner. I diagnosed depression and associated aggression.

A former colleague, who shall remain nameless, objected.

"Dogs don't experience the same mental states and emotions that people do," he said, repeating what back then was the party line.

"Well, how about this?" I answered. "Let's give the dog an antidepression drug and see what happens."

Of course, we know the outcome. Nowadays it is common for veterinarians to prescribe antidepressant drugs designed for people. The rising trend prompted Eli Lilly's animal health division, Elanco, to introduce a chewable beef-flavored version of Prozac for use in dogs.

There's an additional note to sound in the story of Sumo and Jacques Chirac, whose transition out of the French presidency in 2007 was not smooth. Soon after he left office, Chirac faced accusations of financial impropriety and was swept up in a series of official hearings, as he had lost the protection of executive immunity. This must have been extremely stressful—and pets reflect the moods, emotions, and personalities of their owners. The abrupt transition from palace to apartment might have been part of it, but Sumo's awareness of Chirac's distress no doubt contributed to the dog's own upset. Once he was removed from that toxic emotional environment and placed on the farm in the beautiful French countryside, he never bit another person.

When I prescribe medicines for pets, I often hear this same response from clients: "But I thought that was a drug for humans, not animals!"

"Yes," I reply. "But it works the same in animals." I don't point out to them that humans, after all, are animals themselves.

This conversation happens time and time again at our clinic at Tufts. Clients find it most odd when we send them to the local "human" pharmacy to pick up medications for their pets.

"You mean it's exactly the same?"

"Yes, exactly the same." A look of puzzlement and wonder crosses their faces.

If they but stopped a moment and thought about it, they might see how basic is the truth behind the practice. Evolution does not create creatures out of whole cloth. It builds upon what came before. The unity of all life, a concept so beloved by spiritualists, finds its expression in evolutionary biology. All mammals evolved from a mammalian prototype species. There was no need to invent an entirely new nervous systems, new limbic structures, new brain chemistries. What is now here was carried forward from the past.

A veterinarian friend of mine, Dr. Ian Glen, formerly my anesthesiology mentor at Glasgow University, left academia for the commercial world. He joined what was then Imperial Chemical Industries, or ICI, in England. His job was to screen scores of potential injectable or inhalational anesthetics for use in humans, using mice as a model.

One of the experimental, injectable anesthetics Dr. Glen worked on turned out to be exactly what ICI was looking for and later received its now well-known brand name, Propofol. The powerful anesthetic is used in hospitals around the world, but is notorious for being the pop star Michael Jackson's preferred sleep aid. He called it "milk," because that's what it looks like when dissolved in a solubilizing agent for injection. Jackson overdosed and died after his personal physician injected him with it. "More milk," were Michael Jackson's last words.

Of course, any substance on the face of the earth can be abused. Propofol has benefited millions of surgery patients, largely without negative consequences. Punsters have labeled it "milk of amnesia," a real-life version of "the enchanted stem" that puts the "brother mariners" of Ulysses into a trance.

Another acquaintance, Dr. Ross Terrell, developed an inhalation anesthetic, isoflurane, using similar rodent-based experiments. I was sitting next to him at dinner one night.

"Isoflurane has made your company, Anaquest, rich," I said, an excellent opening gambit for any conversation. "Where did the company acquire it from? Who sold it to Anaquest?"

"We didn't buy it," Dr. Terrell replied. "I made it."

Once again, as with Propofol, innovation flowed from laboratory animals directly to the human clinic. It's the kind of thing that happens all the time.

All this arose from experiments first performed on mice. The concept behind my friend's research—and the scientific research of millions of other experiments—is elegant in its simplicity: If it works in a mouse it will work in a person. Of course that would be true. Mice have a brain structure and a central operating system that are very similar to that of humans.

It could almost be called a foundational principle of pharmacology. What happens in animals predicts what will happen in people.

That pill you are placing in your mouth? A white lab rat has been there before you. It's not just drugs for medical conditions, either. Pain medications and anesthetics follow the same path, as do psychoactive drugs for treatment of anxiety and depression. All begin on the drawing board and are then subjected to testing in rodent models before entering human clinical trials. If successful, human clinical trials lead to the licensing of that new drug for use in people. There are no intermediate steps. Like pharmacological boomerangs, some of the drugs then find their way back to use in animals.

Anxiety and aggression have been extensively researched using animals, and through these studies we've also learned about the neurochemical underpinnings of depression as well as how to treat it. As mentioned earlier, the neurotransmitter serotonin and its fluctuations are intimately involved in aggression. In humans and in other animals, serotonin levels fluctuate from dawn to dusk. So do moods, and so does the potential for aggressive responding. We all have a longer fuse in the morning and become more irritable in the late afternoon.

This fluctuation in serotonin levels participates in the so-called circadian rhythm of biological processes, their oscillation over a twenty-four-hour cycle. One other thing that affects the level of serotonin is what happens in the course of life. Certain accomplishments, such as winning a fight, increase serotonin. Defeat has the opposite effect. This result has been clearly demonstrated in monkeys. And, well, monkeys are *us*.

A serotonin experiment that teaches us a lot about ourselves was conducted with vervet monkeys and showed that dominant monkeys had higher serotonin levels than their subordinate cage mates. Dominance was actually associated with lower levels of aggression. Dominant monkeys did not have hair triggers. They didn't need them! They knew they were ultimately in charge. They only responded with aggression when severely pressed.

In one experiment, the researchers removed the most dominant male monkey from the group. This left the others wondering who would be the new boss. A select few of the remaining male monkeys were treated with a serotonin-enhancing drug, Prozac. Those treated always achieved dominant status, became less aggressive, and engaged in more affiliative behaviors, such as mutual grooming. They also attracted more attention from the females.

The reverse was true, too. If the remaining male monkeys were treated with a serotonin antagonist, they always became subordinate, less social, and more aggressive. In other experiments in monkeys, low serotonin levels were shown to be associated with greater risk taking, including lashing out aggressively instead of considering the possible consequences of initiating an altercation.

We know that increasing serotonin decreases aggression. This holds across a range of species, from rodents to humans. Conversely, we know that decreasing serotonin darkens mood and creates impulsivity and aggression. We treat aggression in animals with drugs, but, while no drugs are approved to treat aggression in people, there are plenty of medications that are used off-label to do just that. One of these is clonidine, which works by decreasing the release of the catecholamine norepinephrine from nerve endings in the central nervous system. Its primary use in people is to control blood pressure. But psychiatrists also employ it to control mood.

Reducing the release of catecholamines induces a calmer state of mind. Clonidine finds application in treatment of attention deficit, hyperactivity, mood instability, and aggressiveness, particularly in children. At Tufts, we've long used clonidine as an effective therapy for fear-related aggression in dogs. If we had found that clonidine reduced aggression in dogs before psychiatrists discovered the effect in people, we could have predicted the cross-species antiaggressive applications in human medicine.

When you come to consider the big picture, without animal testing there would not have been any of the major drug developments over the last sixty-five years or so. One huge breakthrough during that period was the discovery of what became known as neuroleptic drugs, later grouped under the antipsychotic class of medicines. Even before its antipsychotic effect was noted, chlorpromazine was the first of these neuroleptic drugs to be studied in humans.

The term neuroleptic comes from the two Greek words: *neuro*, meaning nerve, and *leptikos*, meaning seizure. When chlorpromazine was first given to rodents, the creatures became instantly immobile. Chlorpromazine's antipsychotic effect came from blocking dopamine in the brain. Modification of chlorpromazine led to the development of acepromazine, a tranquilizer used frequently by almost all veterinarians today. Its effect on animals is the same as chlorpromazine on people. ACP, or Ace, as acepromazine is colloquially referred to, calms and sedates agitated animals and is often employed as a preanesthetic, just like chlorpromazine in the original "lytic cocktail."

The first antidepressant, imipramine, was termed a thymolytic, a term that roughly translates into "taking hold of the emotions." In the late 1950s, imipramine was developed directly from chlorpromazine, primarily in an attempt to improve the drug's effectiveness. From there, other antidepressants with the same tricyclic structure were fabricated. The age of psychopharmacology had arrived.

Researchers eventually determined that depression was related to low levels of serotonin in the brain. One of the effects of these early antidepressants was to boost serotonin levels. That realization led to the quest for selective serotonin reuptake inhibitors, the vaunted SSRI class of medicines.

Over the ensuing two decades, many SSRIs were developed. Some were more successful than others. Two of the first to be used were flu-voxamine, later marketed as Luvox in the United States, and citalo-pram, sold as Celexa.

Perhaps the most high profile SSRI of them all, fluoxetine, or Prozac, came on the US market in 1988. Of course, initial testing of Prozac's serotonin reuptake properties occurred in the brains of lab-oratory rodents. Researchers with the Eli Lilly company, such as the scientist David T. Wong and the clinician Jong Sin-Horng, knew SSRIs would have the same effect in human brains. Intuitive, right? It was a no-brainer.

SSRIs are so popular in psychiatry that they are used more than any other class of drug. I once quizzed a professor of psychiatry at Harvard how commonly Prozac-like meds were used in his field.

"Eighty-five percent of the time?" I suggested.

"Oh, more than that," he replied.

In human medicine, SSRIs are used to treat aggression, anxiety, phobias, PTSD, and obsessive compulsive disorder. I will allow you, the reader, who has by now been through almost the whole of this book, to guess which conditions they are used to treat in veterinary medicine. That's right, the same ones.

Regarding human psychiatric drugs and veterinary behavioral drugs, the truth by now is well established. If it will work in animals, it will work in people. Conversely, if it works in people, it will work in animals. As Alexandre Dumas phrased it in *The Three Musketeers*, it's a case of "One for all and all for one."

Epilogue

Hope for Us All

A man is truly ethical when he obeys the compulsion to help all life which he is able to assist, and shrinks from injuring anything that lives.

—ALBERT SCHWEITZER

Aggression is the number one cause of death among pet animals, because patterns of violent behavior all too often lead to them being put down. The following story has one such sad ending.

Ruckus, a four-year old, neutered male, soft-coated Wheaten terrier, more than earned his name. He was brought to Tufts for evaluation after an odd and progressively more serious problem of aggression, coming to me at two years of age. He was the best-looking specimen of the breed that I have ever seen. Spirited and playful when he was first adopted by his owners, Dick and Nancy Tiemer, a handsome couple who clearly loved their dog, Ruckus was, in effect, their surrogate child. During the initial consultation, they both held me with an intense, pleading gaze, as if begging for help.

Dick started off by telling me about their early days with Ruckus.

Apparently, there were a couple of red flags, if only they had been alert enough to see them. The dog's former owner informed Dick and Nancy that Ruckus had been in a crate for about fourteen hours a day. He also gave them a piece of unsolicited advice: "Make sure that you get into bed first at night." At the time, the Tiemers found the suggestion to be somewhat odd, but dismissed it in the midst of their joy over the new puppy. The words would come back to haunt them.

When they got Ruckus home, all was well until evening. Ruckus dashed upstairs ahead of them and commandeered the bed. When they tried to remove him, he held his ground and would not budge. He gave a low growl, indicating the bed was his territory. Other issues became apparent within the first seventy-two hours. Ruckus began to display various ornery traits, including putting up a huge fuss when they tried to crate him. Nancy wondered whether they should return the dog. When they called the previous owner, he refused to take Ruckus back. Dick and Nancy were stuck.

The more modest levels of aggression eventually transformed into barking, teeth-baring, and other threatening displays. Initially, they were able to get control of the aggression using commands like come, sit, and down. The dog would obey, "reluctantly," as the Tiemers put it. After two weeks in the household, Ruckus had already bitten Dick twice.

Their local vet prescribed Prozac for Ruckus, though the med also might have helped the owners. He also suggested that the Tiemers should hire a trainer, who in turn proposed keeping Ruckus confined to one room and the use of baby gates in all doorways. Basically, the pet was to be allowed to move from one area to another only if he was invited to do so, and only if he worked for the privilege by obeying a command. This same trainer also suggested using a muzzle to protect Dick, and an electric shock collar as the ultimate deterrent.

I did not and do not approve of the latter strategy. For me, shock collars always bring to mind a sage comment by Abraham Lincoln:

"Whenever I hear anyone arguing for slavery, I feel a strong impulse to see it tried on him personally." I would advocate the same approach in the case of shock collars.

As bad as the situation was when the Tiemers sought the trainer's help, Ruckus's rage outbursts toward Dick actually increased in frequency and intensity. The local vet recommended euthanasia, saying that Ruckus was "miswired." Dick and Nancy wanted to keep trying. They scheduled an appointment with me at Tufts.

As we sat there in the consulting room, with Ruckus behaving like a perfect angel, they described him as sweet toward strangers, children, and other dogs, as well as with other family members. Apparently, the only difficulty was his aggression toward Dick, but that problem was particularly severe. The Tiemers were willing to euthanize Ruckus, they told me, "If he was truly suffering in some unrelenting way."

I was not so sure Ruckus was suffering. After much discussion, we agreed on a way forward. I upped Ruckus's Prozac dose, gave additional advice about avoiding conflict, and had them discontinue use of the shock collar. I instructed them to make Ruckus work for food and treats. I also suggested another "as needed" medicine, clonidine, that I thought might help at nighttime.

A week later Dick emailed me a progress report. "The week began with more of the same (a Friday night attack and more unpleasantness over the weekend) but then the medications calmed his outbursts dramatically." This sounded encouraging. There was some later backsliding, so I adjusted the dose of both his medications.

For a short period the terrier's behavior improved. We used this window of opportunity to have Dick engage in some counter-conditioning exercises, designed to convert the dog's unpleasant associations with his owner to more favorable ones. Through various upswings and downswings, it became clear that Ruckus continued to be dangerous. It reached a point where Dick had to call Nancy on his cell phone when

PETS ON THE COUCH

he was approaching the house. Nancy would then put the dog away so that Dick could enter his own house without being attacked.

The endgame came about in tragic fashion. Ruckus was outside, tied to a stake while Dick was mowing. With a violent lunge, the terrier tore the stake out of the ground. Tether and stake trailing behind him, he raced toward Dick and attacked him savagely. Eventually, after receiving several nasty bites, Dick managed to pin Ruckus and yell to Nancy to get help.

The Animal Control Officer arrived within minutes. He and his assistants managed to subdue the dog. They applied a transparent plastic muzzle and carried him, still struggling, to their vehicle. Unfortunately, the muzzle was too tight. Ruckus could not breathe. By the time they arrived at the Animal Control Officer's vehicle, the poor terrier had asphyxiated and was gone.

Our victories might fade into the background, but we always remember our defeats. The case of the belligerent terrier remains fresh in my mind even today. It helped reinforce my ongoing determination to discover new ways to treat aggression in pets. No dog should suffer the way Ruckus did, and no animal should experience such a death.

If there is a gleam of light in this terribly dark story, it is the incredible dedication of the Tiemers toward saving their dog. Even in the face of violence and aggression, even with Dick being bitten repeatedly, they did not give up. Their story represents an incredible testimony to human love in the face of adversity, and to the tremendous attachment people can have to their pets.

But there's another reason I've introduced this distressing tale. In the course of this book I've discussed a wide variety of treatments, and many of them employ sophisticated medicines. I have often encountered pushback when I discuss such measures with people. They seem to see a disconnect between our warm and fuzzy friends and the complex molecules we use to treat them. They'd prefer to keep things simple, organic, and straightforward.

When confronted by such an opinion, at a cocktail party, say, or during a lecture, my mind immediately goes to my experiences in the consulting room. There I am confronted by desperate, teary-eyed owners of animals who are suffering. It's very easy to dream about a unicorn world where pets can be healed simply by love and pixie dust. But when a dog or cat is in pain, the stakes turn very serious, very quickly. There are no atheists in foxholes, and there are few starry-eyed herbalists in consulting rooms, holding their suffering pet. In that situation, I challenge anyone to refuse proven scientific measures of relief.

The cases of pet behavior I've discussed in this book range from the serious to the frivolous. Throughout, I've tried to emphasize the principles of One Medicine. But I would not have you think that I am a pill pusher, a Doctor Feel Good who reaches into the medicine cabinet at the first sign of trouble. Nothing could be further from the truth. The majority of the behaviors I encounter can be treated first and foremost with common sense, employing strategies that do not involve pharmacology at all, but changes in an animal's environment and lifestyle, his interactions with you and other beings in his life.

I call them the Four *E*s. They represent the tried and true line of defense against a whole spectrum of unwelcome behaviors and situations.

Enhance. Employ. Exercise. Empathize. That is,

- Enhance your pet's environment with toys, distractions, and stimulating activities.
- Employ a sensible diet to match your pet's needs.
- Exercise your pet daily.
- Finally, and here's the most important one—empathize with your pet's state of being.

In many cases, employing the Four *E*s will have wonderful therapeutic effects on your pet's health. I believe empathizing is vital, since only if we as humans cross the species barrier and try to truly understand a dog, cat, or other animal will we be able to help it. Too often we allow emotional judgments to color our relations with animals.

I can hear the pet owner's voice in my head. "I am so angry with my dog. She refuses to stop barking at strangers! It's driving me nuts. I'm so embarrassed. I try to shush her, but it's like she's doing it just to annoy me."

Well, she's not doing anything just to annoy you. Her behavior has deep-seated reasons. Aggression against strangers might indicate deeper behavioral issues.

"Oh, okay, so you're telling me, Dr. Dodman, that my dog might be fearful of strangers because of possible mistreatment in early puppyhood? And that there are measures we can take to address the issue? So I don't have to take her back to the pound after all?"

Empathy helps where judgment fails. But empathy depends in part on identification. We can't empathize if we feel a creature is somehow alien, different, or separated from us by insurmountable barriers. My emphasis has always been upon our commonality with our pets, not our differences.

Let's return to my initial considerations. Are animals like us and are we like them? Do they have intelligence? Do they have emotions? Can they suffer as we sometimes do?

Arguments about animal intelligence have been going on for a very long time. One of the most well-established reasons for differentiating nonhuman animals from people was that animals supposedly did not employ tools. Tool creation and use are what gave humans the edge. That particular line in the sand got erased in the 1960s, when pioneering primatologist Jane Goodall stunned the scientific community by showing that chimpanzees used tools in a variety of situations. The first discovery she made involved chimps using long twigs to fish tasty termites out of the mounds. Chimps also fashioned "sponges" of crushed leaves on sticks in order to soak up rainwater to drink. They utilized sticks and rocks to smash fruit with hard shells and to intimidate others, too.

In response to these discoveries, Dr. Louis Leakey wrote: "Now we must redefine tool, redefine man, or accept chimpanzees as humans."

Other tool-using animals were soon identified. Monkeys, elephants, dolphins, sea otters, mongooses, American badgers, ravens, finches, alligators, and crocodiles were all caught red-handed—or flippered, or beaked, or clawed—using tools.

Given Leakey's choice, the purists chose to redefine man. No, no, humans weren't the tool-using animal, after all. We were now, instead, defined as the language-using animal.

Soon enough, this distinction began to crumble as well. Apes in the wild use complex vocalizations. In captivity, gorillas and chimpanzees have been taught to use sign language, and African gray parrots employ large vocabularies.

Okay, okay, so man is not the language-using animal. We're . . . we're . . . something else special! Memory! What about memory?

Quite recently there was a report that surprised the media, showing that chimps and orangutans could remember things for years. This finding indicated that the memory function of great apes resembled our own. Experiments with King, a western lowland gorilla at Florida International University, established that he could recall who fed him what foods even when his caretakers had forgotten. And as was shown in a remarkable demonstration, chimpanzees can actually triumph over humans in short-term memory trials (http://on.fb.me/1OTZo9n).

As soon as one distinction fell, another was erected in its place. The bar always had to be set a little higher.

Just hold on, we'll come up with something to distinguish us from the beasts!

But what if animals actually do think about what they are doing? What if they have feelings and emotions? What if they are more like us than was previously thought?

The four species of great apes are currently locked in a legal fight in courts around the world, their advocates pursuing legal personhood for gorillas, orangutans, bonobos, and chimpanzees. Steven M. Wise of the Nonhuman Rights Project has filed briefs for captive chimpanzees in New York State and elsewhere, hoping to establish their rights to freedom under the writ of habeas corpus.

To invoke another example, India's Ministry of Environment and Forests now accepts that dolphins and other cetaceans are highly intelligent and sensitive animals. Specifically, the ministry agrees with various scientists who consider that because of unusually high intelligence displayed by dolphins, they should be seen as "non-human persons." As such, dolphins would enjoy their own specific rights. It would be recognized as morally unacceptable and legally untenable to keep them captive for entertainment purposes.

A team from the University of St. Andrews in Scotland discovered that dolphins use signature whistles to communicate with each other, much as we use names. The team believes the dolphins are acting like humans, in that when they hear their name, they answer. This ability implies that dolphins are self-aware and fully cognizant of other dolphins around them. That dolphins are self-aware may sound obvious to the average person. Such a statement would be rejected by extreme behavioral scientists. Such folks always clamor for proof of anything cerebral and non-reflexive in animals. With dolphins, they now have it.

The apparent lack of language among animals has always been a barrier to scientists accepting them as intelligent entities. There are people who believe that language is necessary for thought. They believe that without a fluent vocabulary of interconnecting words, deliberation and reflection are simply not possible. This argument has largely been invalidated by accounts of people who have temporarily lost language as a result of some cerebral injury, such as a stroke. Certainly, life is different without language. But thought is not lost.

Prelingually deaf people who have never heard or uttered a single word still think. They sometimes gather in groups, miming their thoughts to each other as the only way they have of communicating. It takes a bit longer, but they succeed. Human babies have been shown to communicate long before their spoken language skills are developed. And Temple Grandin has shown that at least some autistic people think in pictures.

In the 1850s, Charles Darwin wrote a fundamental truth. Humans are animals. We are simply part of a grand tableau of creation. We are not its purpose. Darwin believed that our emotions and intellect, and not just our bodily form, evolved from a common primate ancestor. He suggested that we can best understand ourselves by studying the psychological, as well as physical, steps in evolution. That's what I have been attempting to address in this book, by considering the psychological, psychiatric and behavioral parallels that exist between people and pets.

Much of our own mental life is automatic, unconscious, and outside the control of our reason or will. Certainly we can stop and think. But we also operate on autopilot much of the time. Nonhuman animals also appear to stop and think, too, though we can't ask them what they are thinking about. Experiments have been designed to examine thought processes in animals. All are rather simple and equivocal in terms of what they actually mean.

One experimental setup had dogs exploring a T-maze, with a reward at one of the limbs of the T. When a dog reached the point of divergence of the limbs of the maze, he hesitated before committing. This hesitation, which can be measured, was believed to represent the dog vacillating over which way to go.

At a recent meeting I attended, I heard about a similar situation that occurred with cats in a T-maze. The researcher observed the hesitation at the T-junction. She apologized to the erudite audience, which was made up largely of behavioral scientists, about the possible anthropomorphic interpretation.

"I realize you are all behaviorists, so I'm almost afraid to say what I think this means," she said. "It does look for all the world like the cats are confused and trying to think which limb would be best to explore."

Despite Darwin, despite Goodall, despite Temple Grandin, and so many others, we still find ourselves having to apologize in scientific circles for ascribing the power of thought to animals. It now requires a head-in-the-sand approach to deny that the similarities between mammals far outweigh the differences.

Today mammals, tomorrow the avian kingdom. Birds, especially psittacines, seem very much cognitively attuned. You have to live with birds to fully appreciate their intelligence. My Amazon parrot, whom I nicknamed Green Jungle Chicken, convinced me that animals can have skills far beyond what is normally assumed.

This bird was more like a dog than some dogs I know. Chicken, as we called her, flew free in our residence, and had many humanlike qualities. She had an incredible vocabulary, wolf-whistled when anyone was naked in the shower, greeted me on my return home by flying onto my shoulder, and flew up to the upstairs landing, anticipating that I was heading to bed. Chicken liked her coffee with milk and one sugar. She would drink it from a spoon held in her large gray claw. There's an instance of tool use for you!

Living with Chicken made me think of Pippit and my mother's wild flock of birds. Gwen Dodman used to have a parrot, too, an African gray called Polly. Polly would always ask for the remains of the apple before anyone was finished with the fruit. "Polly wants the core," she'd remind us, squawking fitfully. When my father would raise his hand and pretend to be about to strike my mother, Polly would also fly to her defense, attacking my father by latching onto his rear end with her beak.

One fascinating story involves a humpback whale showing gratitude after being freed from entanglement in fishing nets. The divers worked for hours to free the huge mammal. They finally achieved their goal. Once free, the whale dived off into the murky depths. Minutes later, one of the divers saw the huge whale heading up straight for him at top speed. As the whale got close, it slowed down. The fifty-ton beast, the size of a school bus, came to rest a few inches from the diver's chest.

The whale gently nudged the human in the chest a few times. It then raised its head out of the ocean to stare at the diver for more than half a minute. For the diver, it represented a haunting experience. "I'll take it to my grave," he said. The whale then went on to the next diver, and the next, until it had made the full round of all the wet-suited Samaritans.

No doubt an extreme behaviorist—that is, one bound by Morgan's Canon, with no interpretation of animal behavior permitted—would try to find some reflex explanation of the creature's actions. To me, the truth is clear. The whale was clearly showing gratitude.

All the conditions I have described in this book involve animals being to some degree aware and sentient. They sometimes become anxious or even angry in this hectic modern-day life. They reflect similar mental issues to the ones we ourselves may experience.

Some of the behaviors in question have "face validity" for their equivalent human conditions—that is, the actions look similar. Acral lick dermatitis, for example, resembles its human hand washing equivalent. Almost all animal equivalents of human behavioral conditions have "predictive validity" for the corresponding human condition. It is quite possible to predict that what works in humans to alleviate an anxious disorder will work in animals, and vice versa. The anti-OCD strategy of using an NMDA blocker was first tested by us in mice, dogs, and horses before it was ever found to work in people.

These examples provide clear documentation of this predictive validity. The final element of sameness is "construct validity"—meaning finding similar genes, similar structural homology, and similar biomarkers. Though this last element of proof is incomplete for most animal behavior problems, the net is closing and the evidence is mounting.

Face validity, predictive validity, and construct validity. Game, set, and match. It's pretty much everything we need to know to establish true homology with the equivalent human conditions.

Here's just a few examples that we've delineated in the preceding chapters:

- The neural cadherin gene's involvement in canine and human compulsive behavior.
- Fragile X syndrome in bull terriers and in some autistic children.
- Elevation of plasma neurotensin in spinning bull terriers, which also occurs in autistic children.
- A serotonin gene's involvement in aggression in Golden retrievers, with serotonin and aggression inversely related in man and animals.
- Brain centers underlying aggression are the same across mammalian species.
- Brain changes in compulsive Doberman pinschers are the same as reported in humans with OCD, especially hoarders.
- Amyloid plaques and tau tangles occur in canine cognitive dysfunction, as they do in people with Alzheimer's disease.

Soon it will be very difficult for the naysayers to deny the parallels between animals and people. As Darwin realized, though animals aren't people, people are animals.

Certainly, no doubt about it, humans have the edge cognitively. We possess the mental equivalent of a mainframe computer atop our shoulders, while the animals must make do with the less sophisticated but functionally similar Commodore 64 version. Animals clearly are capable of thought, can remember events for a long time, are aware of time passing, can become upset, angry, and sad. They have primary and secondary emotions and are capable of suffering and being anxious.

As humans excel in some areas, nonhuman animals beat us hands down in others. Animals are particularly good at tailoring their facilities according to their ethological niche.

- Dogs are better than us at reading body language. This is a skill they need in the pack. Dogs are also better than us at making mental maps. That's a skill that facilitates hunting and getting home after a long trek across their vast home range.
- Cats have outstanding balance and excel at kinesthetic intelligence. They are better at orienting their bodies in space than the greatest human gymnasts. This talent suits them well in their arboreal pursuits and is useful for snagging birds on the wing.
- Horses see separate visual fields and can distinguish what they see on either side of their body. That's a useful gift for a prey animal as it helps them to see all around their body at the same time.

Consider poor humans, with our lesser vision and auditory ability, weak olfactory capability, and poor body language skills. What we do have that they don't have, though, to suit our own biological agenda, is a highly developed brain and a specialized version of a particular gene, called FoxP2, that allows for the proper development of speech and language.

In the days when mass immigration occurred into the United States via Ellis Island, people were judged according to parochial standards. Many were deemed mentally challenged, imbecilic, idiots, or morons. Such judgments arose because the applicants for entry could not answer questions about, for example, American civics or jurisprudence, in tests posed in a language that they did not fully comprehend. But at times the immigrants seemed smarter than the examiners.

"If you would use a broom to clean stairs, would you clean from the bottom up or the top down?" asked an Ellis Island officer of a female immigrant.

"I didn't come to America to sweep stairs!" she replied.

We must be careful when judging animals by our own standards, so as not to make the mistake of the Ellis Island authorities.

Fortunately, using modern imaging techniques, it is now possible to peer inside an animal's brain. We can see it think. We can gauge its response to visual or auditory cues in a sophisticated way, right down to identifying the brain regions activated. Using such techniques, along with identification of the pathways activated by the relevant genes, we should soon be able to scientifically substantiate everything I have suggested as true in this book.

In identifying psychological and psychiatric conditions that are common in separate species, it was never my intention to reduce humans to the level of beasts. Rather, I'd seek to raise our appreciation of the complex interior lives and mental abilities of our pets. We need to more fully appreciate their struggles. We are all animals in this life together, and we should respect the right of nonhuman creatures to be treated kindly and respectfully.

Acknowledgments

I would like to thank my super "hands-on" editor, Leslie Meredith, for steering me in the right direction with this book; my supremely patient agent Paul Bresnick for his steadfast advice and support; and Gil Reavill, who helped get the final manuscript into such good shape. I would also like to thank all the clients who allowed me to use stories about their pets in this book. You are all wonderful!

Bibliography

Peer-Reviewed Articles

Borns-Weil, Stephanie, Ashley Smith, Christine Emmanuel, Jami Longo, Nicole Cottam, and Nicholas Dodman. "A case-control study of compulsive wool-sucking in Siamese and Birman cats (n=204)." *Journal of Veterinary Behavior* 10, no. 6 (2015): 243–48.

Cornwall, E., E. Brown, K. Damiani, and N. H. Dodman. "Thunderstorm phobia in dogs: A survey of 69 cases." *Journal of the American Animal Hospital Association* 37 (2001): 319–24.

DeNapoli, J. S., N. H. Dodman, L. Shuster, W. M. Rand, and K. L. Gross. "Effect of dietary protein content and tryptophan supplementation on dominance aggression, territorial aggression, and hyperactivity in dogs." *Journal of the American Veterinary Medical Association* 217, no. 4 (2000): 504–8.

Dodman N. H. "Pharmacological treatment of behavioral problems in cats." *Veterinary International* 6 (1994): 13–20.

———. "Prozac shows promise in treating behavior problems." *Veterinary Medicine* 92, no. 4 (Apr. 1997): 318–19.

Dodman, N. H., A. Weissman, and W. Walker. "Animal behavior case of the month." *Journal of the American Veterinary Medical Association* 220, no. 5 (2002): 604–7.

Dodman, N. H., and B. Olivier. "Search of animal models for obsessive compulsive disorder." *CNS Spectrums* 1, no. 2 (1996): 10–13.

Dodman, N. H., and D. Arrington. "Animal behavior case of the month." *Journal of the American Veterinary Medical Association* 217, no. 10 (2000): 1468–72.

Dodman, N. H., and L. Shuster. "Pharmacologic approaches to managing behavior problems in small animals." *Veterinary Medicine,* Oct. 1994, 960–69.

Dodman, N. H., and N. Cottam. "Animal behavior case of the month." *Journal of the American Veterinary Medical Association* 225, no. 9 (2004): 1339–41.

———. "The effectiveness of the anxiety wrap in the treatment of canine thunderstorm phobia: An open-label trial." *Journal of Veterinary Behavior: Clinical Applications and Research* 8 (2013).

Dodman, N. H., E. A. Billingham, and A. A. Moon-Fanelli. "Animal behavior case of the month." *Journal of the American Veterinary Medical Association* 227, no. 2 (2005): 228–32.

Dodman, N. H., J. A. Normile, L. Shuster, and W. Rand. "Equine self-mutilation syndrome—a series of 57 cases." *Journal of the American Veterinary Medical Association* 204, no. 8 (1994): 1219–23.

Dodman, N. H., K. A. Miczek, K. Knowles, J. G. Thalhammer, and L. Shuster. "Phenobarbital responsive episodic dyscontrol in dogs." *Journal of the American Veterinary Medical Association* 201 (1992): 1580–83.

Dodman, Nicholas H., Elinor K. Karlsson, Alice Moon-Fanelli, Marzena Galdzicka, Michele Perloski, Louis Shuster, Kerstin Lindblad-Toh, and Edward I. Ginns. "A canine chromosome 7 locus confers compulsive disorder susceptibility." *Molecular Psychiatry* 15 (2010): 8–10.

Dodman, N. H., K. Knowles, L. Shuster, A. A. Moon-Fanelli, A. Tidwell, and C. L. Keen. "Behavioral changes associated with suspected complex partial seizures in Bull Terriers." *Journal of the American Veterinary Medical Association* 208 (1996): 688–91.

Dodman, N. H., L. Shuster, S. D. White, M. H. Court, D. Parker, and R. Dixon. "The use of narcotic antagonists as therapeutic agents to modify stereotypic self-licking, self-chewing and scratching behavior in dogs." *Journal of the American Veterinary Medical Association* 193 (1988): 815–19.

Dodman, N. H., L. Shuster, M. H. Court, and J. Patel. "Investigation into the use of narcotic antagonists in the treatment of a stereotypic behavior pattern (crib-biting) in the horse." *American Journal of Veterinary Research* 48 (1987): 311–19.

Dodman, N. H., L. Shuster, G. Nesbitt, A. Weissman, W.-Y. Lo, W. Chang, and N. Cottam. "The use of an NMDA receptor antagonist, dextromethorphan, to treat repetitive self-directed scratching, biting, or chewing in dogs with allergic dermatitis." *Journal of Veterinary Pharmacology and Therapeutics* 27, no. 2 (2004): 99–104.

Dodman, N. H., L. Shuster, G. Patronek, and L. Kinney. "Pharmacologic treatment of equine self-mutilation syndrome: A pilot study." *Journal of Applied Research in Veterinary Medicine* 2, no. 2 (2004).

Dodman, N. H., L. Shuster, M. H. Court, and J. Patel. "The effects of nalmefene on self-mutilative behavior in a stallion." *Journal of the American Veterinary Medical Association* 192 (1988): 1585–86.

Dodman, N. H., L. Shuster, M. H. Court, and S. D. White. "Behavioral effects of narcotic antagonists." *Journal of the Association of Veterinary Anaesthetists* 15 (1988): 56–64.

Dodman, N. H., N. Cottam, L. Aronson, and W. J. Dodds. "The effect of thyroid replacement in dogs with suboptimal thyroid function on owner-directed aggression: A randomized, double-blind, placebo-controlled clinical trial." *Journal of Veterinary Behavior* 8 (Feb. 2013).

Dodman, N. H., P. A. Mertens, and L. P. Aronson. "Animal behavior case of the month." *Journal of the American Veterinary Medical Association* 207, no. 9 (1995): 1168–71.

Dodman, N. H., R. Bronson, and R. Gliatto. "Tail chasing in a Bull Terrier." *Journal of the American Veterinary Medical Association* 202 (1993): 758–60.

Dodman, N. H., R. Donnelly, L. Shuster, P. Mertens, W. Rand, and K. Miczek. "Use of fluoxetine to treat dominance aggression in dogs." *Journal of the American Veterinary Medical Association* 209, no. 9 (1996): 1585–87.

Dodman, Nicholas H., Edward I. Ginns, Louis Shuster, Alice A. Moon-Fanelli, Marzena Galdzicka, Andrew Borgman, Lisa Kefene, Jiashun Zheng, Alison L. Ruhe, and Mark W. Neff. "Genomic risk for severe canine compulsive disorder, a dog model of human OCD." *International Journal of Applied Research in Veterinary Medicine* 14, no. 1 (2016): 1–18.

Flannigan, G., and N. H. Dodman. "Risk factors and behaviors associated with separation anxiety in dogs." *Journal of the American Veterinary Medical Association* 219, no. 4 (2001) 460–66.

Grandin, T., N. H. Dodman, and L. Shuster. "Effect of naltrexone on relaxation induced by lateral side pressure in pigs." *Pharmacology, Biochemistry and Behavior* 33 (1989): 839–42.

Hart, B. L., R. A. Eckstein, K. L. Powell, and N. H. Dodman. "Effectiveness of buspirone on urine spraying and inappropriate urination in cats." *Journal of the American Veterinary Medical Association* 203, no. 2 (1993): 254–58.

Maurer, B. M., and N. H. Dodman. "Animal behavior case of the month." *Journal of the American Veterinary Medical Association* 231, no. 4 (2007): 536–39.

———. "Use of memantine in treatment of canine compulsive disorders: A preliminary, clinical trial." *Journal of Veterinary Behavior: Clinical Application and Research* 4 (2009): 118–26.

Mertens, P. A., and N. H. Dodman. "Medical therapy for canine acral lick dermatitis in dogs." *Kleintierpraxi* 41, no. 5 (1996): 327–37.

Moon-Fanelli, A. A., and N. H. Dodman. "Description and development of compulsive tail chasing in terriers, and response to clomipramine treatment." *Journal of the American Veterinary Medical Association,* Apr. 1998: 1252–57.

Moon-Fanelli, A. A., N. H. Dodman, and N. Cottam. "Clinical presentation of blanket and flank sucking in Doberman Pinschers." *Journal of the American Veterinary Medical Association* 231, no. 6 (2007): 907–12.

Moon-Fanelli, A. A., T. R. Famula, N. Cottam, and N. H. Dodman. "Characteristics of compulsive tail chasing and associated risk factors in Bull Terriers." *Journal of the American Veterinary Medical Association* 238, no. 7 (2011): 883–89.

Moya, P. R., N. H. Dodman, K. R. Timpano, L. M. Rubenstein, Z. Rana, R. L. Fried, L. F. Reichardt, et al. "Rare missense neuronal cadherin gene (*CDH2*) variants in specific obsessive-compulsive disorder and Tourette disorder phenotypes." *European Journal of Human Genetics,* Jan. 2013.

Ogata, N., and N. H. Dodman. "The use of clonidine in the treatment of fear-based behavior problems in dogs: An open trial." *Journal of Veterinary Behavior* 6 (2011): 130–37.

Ogata, Niwako, Timothy E. Gillis, Xiaoxu Liu, Suzanne M. Cunningham, Steven
B. Lowen, Bonnie L. Adams, James Sunderland-Smith, et al. "Brain structural
abnormalities in Doberman Pinschers with canine compulsive disorder." *Progress
in Neuro-Psychopharmacology and Biological Psychiatry* (2013), doi:1016/j
.pnpbp.2013.04.002.

Rendon, R., L. Shuster, and N. H. Dodman. "The effect of the NMDA receptor
blocker, dextromethorphan, on cribbing in horses." *Pharmacology Biochemistry and
Behavior* 68 (2001): 49–51.

Sawyer, L. S., A. A. Moon-Fanelli, and N. H. Dodman. "Psychogenic alopecia in cats:
11 cases (1993–1996)." *Journal of the American Veterinary Medical Association* 214,
no. 1 (1999): 71–74.

Schneider, B. M., N. H. Dodman, D. Faissler, and N. Ogata. "Clinical use of an herbal
extract (Huperzine A) to treat putative complex partial seizures in a dog." *Epilepsy
& Behavior* 15 (2009): 529–34.

Shuster, L., N. H. Dodman, T. D'Allesandro, and S. Zuroff. "Reverse tolerance to the
stimulant effects of morphine in horses." *Journal of Equine Veterinary Science* 4
(1984): 233–36.

Stein, D. J., N. H. Dodman, P. Borchelt, and E. Hollander. "Behavior disorders in
veterinary practice: Relevance to psychiatry." *Comprehensive Psychiatry* 35, no. 4
(1994): 275–85.

Stein, D. J., N. H. Dodman, and A. A. Moon-Fanelli. "Cats and obsessive-compulsive
disorder." Letter. *South African Medical Journal* 86, no. 12 (1996): 1614–15.

Stewart, S. E., E. A. Jenike, D. M. Hezel, D. E. Stack, N. H. Dodman, L. Shuster,
and M. A. Jenike. "A single-blinded case-control study of memantine in severe
obsessive-compulsive disorder." *Journal of Clinical Psychopharmacology* 30, no. 1
(2010): 34–39.

Tsilioni, Irene, Nicholas Dodman, Anastasia I. Petra, Ana Taliou, Konstantinos Francis,
Alice Moon-Fanelli, Louis Shuster, and Theoharis C. Theoharides. "Elevated serum
neurotensin and CRH levels in children with autistic spectrum disorders and
tail-chasing Bull Terriers with a phenotype similar to autism." *Translational Psychiatry* 4, e466; doi:10.1038/tp.2014.106. Published online Oct. 14, 2014.

Uchida Y., N. H. Dodman, J. DeNapoli, and L. Aronson. "Characterization and
treatment of 20 canine dominance aggression cases." *Journal of Veterinary Medical
Science* 59, no. 5 (Jan. 1997): 397–99.

Uchida, Y., N. H. Dodman, and D. DeGhetto. "Separation anxiety and stereotypy in a
captive American black bear." *Journal of the American Veterinary Medical Association*
212, no. 3 (Feb. 1998): 354–55.

Wald, R., N. Dodman, and L. Shuster. "The combined effects of memantine and
fluoxetine on an animal model of obsessive compulsive disorder." *Experimental and
Clinical Psychopharmacology* 17, no. 3 (2008): 191–97.

Wrubel, K. M., N. H. Dodman, A. A. Moon-Fanelli, and L. S. Maranda. "Interdog household aggression: 38 cases (2006–2007)." *Journal of the American Veterinary Medical Association*, 238 (6) (2011): 731–40.

Books/Book Chapters

Dodman, N. H. "A Dysfunctional Animal Model of OCD." In *Concepts and Controversies in Obsessive-Compulsive Disorder*, edited by J. S. Abramowitz and Arthur C. Houts. Hingham, MA: Kluwer Academic Publishers, 2005.

———. "Geriatric Behavior Problems." In *Psychopharmacology of Animal Behavior Disorders*, edited by N. H. Dodman and L. Shuster. Malden, MA: Blackwell Scientific, 1998.

———. "Pharmacologic Treatment of Aggression in Veterinary Patients." In *Psychopharmacology of Animal Behavior Disorders*, edited by N. H. Dodman and L. Shuster. Malden, MA: Blackwell Scientific, 1998.

———. "Veterinary Models of Obsessive-Compulsive Disorder." In *Obsessive-Compulsive Disorders: Theory and Management*, edited by Michael Jenike. 3rd ed. St. Louis, MO: Mosby, 1998.

Dodman, N. H., and L. Shuster. "Animal Models of Obsessive-Compulsive Disorder: A Neurobiological and Ethological Perspective." In *Concepts and Controversies in Obsessive-Compulsive Disorder*, edited by J. S. Abramowitz and Arthur C. Houts. Hingham, MA: Kluwer Academic Publishers, 2005.

———. "Spontaneously Occurring Animal Models of OCD." In *Obsessive-Compulsive Disorder: Phenomenology, Pathophysiology and Treatment*, edited by Christopher Pittenger. New York: Oxford University Press, Summer 2016.

Mertens, P. A., and N. H. Dodman. "Pharmacologic Treatment of Fear and Anxiety in Animals." In *Psychopharmacology of Animal Behavior Disorders*. Malden, MA: Blackwell Scientific, 1998.

Moon-Fanelli, A. A., N. H. Dodman, and R. O'Sullivan. "Veterinary Models of Compulsive Self-Grooming: Parallels with Trichotillomania." In *Trichotillomania: New Developments*, edited by D. Stein. Washington, DC: American Psychiatric Press, 1999.

Shuster, L., and N. H. Dodman. "Basic Mechanisms of Compulsive and Self-Injurious Behavior." In *Psychopharmacology of Animal Behavior Disorders*, edited by N. H. Dodman and L. Shuster. Malden, MA: Blackwell Scientific, 1998.

Index

A

acepromazine (Ace or ACP), 247
acetylcholine, 196
acetyl L-carnitine, 190
acral licking (acral lick dermatitis), 24,
    57–60, 68. *See also* lick granuloma
  as anxiety-driven compulsive disorder,
    210
  medications to treat, 58–59, 60, 72
  overgrooming and, 63
  owner's frustration at, 59
  Rapoport experiment, 60–61
  as a stereotypy, 59
  treatment, ineffective, 57, 59–60
ADHD (attention deficit hyperactivity
    disorder), 88, 199–203
  canine ADHD, 202, 203
  dog breeds prone to, 202
  stimulants to treat, 200–201, 203
adrenaline, 36, 37
Afghan hounds, 188–89, 216, 218, 226
aggression, 254. *See also* predation
  animal experiments and, 245–48
  animals provoked to self-defense, 120
  atypical causes, 143–47
  behavioral modification for, 143, 150
  bizarre, in dogs, 1–2, 4–5
  brain centers and, 260
  bull terriers and, 2, 82, 85, 97
  castration to reduce, 122
  in cats, 95, 96, 127, 129–30, 148,
    205–6, 231
  dealing with, 119–20
  depression and, 241–42
  deprived upbringing, adverse events,
    and, 125
  dog bites, 122, 127–29, 130, 134–35,
    249–52
  euthanasia and, 119, 226, 249
  fear aggression, 125–27, 246, 254
  genetic component, 143
  "idiopathic," 4
  instrumental aggression, 131–33
  lack of exercise and, 117–18
  lack of recognition as a condition, 150
  as leading behavioral problem in all
    species, 119
  male and intermale aggression,
    121–24
  maternal aggression, 124–25
  medications for, 136, 139, 143–52,
    246, 250, 251
  misdirected aggression, 128–31
  in monkeys, 146
  Moyer's taxonomy of, 121
  predatory aggression, 137–40
  prolactin and, 125
  protein in the diet and, 118, 128, 131,
    204
  RBD or REM Behavior Disorder in
    humans and, 220
  rescue dog and fear-based, 115–19
  seizure disorders as cause, 2–3, 4, 5, 88,
    95–96, 98
  serotonin and, 144–52, 245–46, 260
  story of Ruckus, 249–52
  territorial aggression, 133–36
  testosterone and, 121–22, 125

aggression (*cont.*)
  thyroid problems and, 227–28, 229, 230, 231
  treatment, 2–3, 130–31
agoraphobia, 159, 176, 182–85
  of cow Cassie, 182–83
  defined, 182
  in dogs, 183–84
  in horses, 184
  treatment, 184–85
Alex (famous parrot, feather plucking by), 66–67
altruism in animals, 9–10
Alzheimer's disease, 187–98, 225
  beta-amyloid and senile plaques, 194, 196, 197–98, 260
  canine version, 187–90, 195, 197–98
  in cheetahs, 195, 197
  dietary supplements for, 190
  diet for, 189–90
  DISHA (symptoms of), 188, 193
  feline version, 191–92
  in humans, 192–94, 198
  medications for, 188–89, 195–97
  parallels between and human and animal versions, 194–95, 260
  social life and, 190
  tau tangles and, 194, 196, 260
  as type 3 diabetes, 195
American College of Veterinary Behaviorists, 12, 39
American Veterinary Medical Association, 12
amitriptyline, 139, 145, 203, 221, 234
amygdala, 36
  fear response and, 36, 38, 157
  genetic predisposition to PTSD and, 39
  seizure disorder in, 91
Anafranil. *See* clomipramine
Anaquest, 244
animals
  altruism in, 9–10
  anatomical similarities across species, 13, 49–51
  anxiety in, 156, 158, 261
  author's joint study of "reverse tolerance," 21

"bad behavior" as leading cause of euthanasia, 7–8
behavioral issues and psychological conditions that also appear in humans, 6, 8–9, 11–12, 13, 49–50, 157, 165, 182–203, 210, 225, 259, 262 (*see also* ADHD; aggression; Alzheimer's disease; anxiety; autism; depression; obsessive compulsive disorder; phobias; PTSD; seizure disorders; sleep problems; thyroid problems; *specific behaviors*)
complex vocalizations by, 255
cruelty toward, 48–49
damage done by outmoded ideas about, 7
deep animal-human emotional link, 27
Descartes and animals as automatons, 10, 12–13, 209
experiments on thought processes of, 257–58
face validity, predictive validity, and construct validity of animal behaviors with corresponding human condition, 259–60
gratitude shown by, 259
human exceptionalism approach to, 7, 16, 26, 30, 220
humans as, 257, 260
intelligence in, 6, 9, 51, 66, 254–55
learned helplessness in, 239–40
legal personhood for great apes, 256
mechanistic theory and, 7, 156
memory and, 255, 261
misjudging, 262
moral behavior in, 27
Morgan's Canon and, 10–11, 13, 16, 156, 165, 259
nonhuman, abilities surpassing human, 261
as nonhuman persons, 256, 257
One Medicine approach, 8
pain felt by, 10, 48
"pathetic fallacy" and, 11, 16–17
predisposition to various fears, 155
psychosomatic conditions in, 209–12
range of emotions in, 6, 7, 16, 51–52, 157, 255, 261

"secondary emotions" in, 9, 51–52, 157
sense of self, or "theory of mind" and, 51–52, 157
sentience in, 47, 256–59, 261
socialization of, 127–28, 162
suffering by, 6, 15, 209, 261
tool-using by, 254–55
veterinary profession and, 12–13
Anthropologist from Mars, An (Sacks), 167
antianxiety drugs, 8, 66, 83, 92, 136, 162, 164–65, 175
anticonvulsants, 64, 83, 88, 89, 91, 92, 95, 96, 98, 136
antidepressants, 109–10, 139, 143, 149, 200, 234
  first drug used as, 247
  MAOIs, 188
  side effects, 144–45
  SSRIs, 165, 247–48 (*see also* Prozac)
  tricyclic, 144–45, 149, 221, 222
anti-obsessional medication, 64, 66, 68
antipsychotic medication, 111, 148–49, 247
dyskinesias and, 148–49
anxiety, 7, 8, 153–71, 241, 261
  animal experiments and, 245–48
  anxiety-based territorial aggression, 135
  appetite and, 210
  brain and, 157
  bulimic gorilla and, 66
  compulsive overgrooming and, 69
  dog behavioral symptoms of, 157–58
  expressions of, 154, 156
  fear vs., 154–55
  generalized anxiety disorder (GAD), 163–64
  health consequences of, 212
  hypothyroidism and, 226–27
  IBS and, 225
  in Iraq War veteran dog, 32–33
  medications for, 8, 66, 83, 92, 136, 162, 164–65, 175
  nocturnal anxiety, 201, 215, 216–18
  nocturnal anxiety example, in dog, 28–30
  as objectless, 155
  OCD and, 61, 69, 154, 164
  owner's, transmitted to pet, 118
  pain and, 208
  panic attacks and, 158–59

psychogenic alopecia in cats and, 67
separation anxiety, 27–30, 42, 159–62, 199, 200
situational anxiety, 154–55
social anxiety, 162–65
stranger anxiety, 153–54, 163, 254
tail chasing and, 63–64
Aricept, 196
Asperger's syndrome, 166
attention deficit hyperactivity disorder (ADHD). *See* ADHD
autism, 166
  biomarker for, 87
  canine autism, 85–88, 260
  CDH8 (gene) and, 87
  endorphins and, 171
  environmental factors and, 88
  flavonoids for, 87
  genetic component, 87
  seizure disorders with, 85, 88
  squeeze machine and, 167, 168
Awakenings (Sacks), 148

B

baboons, 211
Baby (cat with hyperthyroidism), 230–31
Bailey (dog with fear-aggressive behavior), 115–19, 128
barking, 52, 117, 125, 157, 164, 216, 250, 254
  compulsive behavior and, 63
  lack of, 164
  separation anxiety and, 160, 161
  territorial alarm barking, 117, 134
Baypath Humane Society of Hopkinton, Mass., 41–43
beagles, 199–201
Bear (dog with narcolepsy), 222
bears, 65
behavioral problems. *See also* aggression; *specific problems*
  ADHD and, 199–203
  brain tumor and, 205–7
  hypothyroidism and, 226
  lissencephaly and, 207
  liver disorder and, 203–5
  medical underpinnings for, 201
  pain and, 208

behavioral problems (*cont.*)
  psychosomatic conditions and, 209–12
  rabies and, 205
  separation anxiety and destroying
    objects, 160, 199
Belgian schipperke, 222
Benny (dog with seizures set off by
  sunlight), 91, 98
Bentham, Jeremy, 15
Bernese mountain dog, 92
Bernie (dog with seizure disorder), 92–93
beta-blockers, 36–37, 147, 149, 175
bexarotene, 197
Birchland Park Middle School, 112
birds. *See also* parrots
  cognition in, 258
  feather plucking in, 66–67, 68, 210
  preening behavior, 66
blanket-sucking, 62, 64, 70, 87
Blizzard (dog who chases his tail), 82, 88
Boodil, My Dog, 84
border collie, 187–90
Boston terriers, 218
boxers, 218
Boy Who Couldn't Stop Washing, The
  (Rapoport), 60
brachycephalic airway syndrome (BAS),
  218–19
brain. *See also* Alzheimer's disease;
    amygdala; seizure disorders
  aggression and, 260
  aging and, 197
  basal ganglia, 24
  cancerous tumor, 205–7
  compulsions based in specific areas, 24
  compulsive behaviors and, 75, 78–79
  data on fear and anxiety, 156–57
  EEGs for dogs, 5, 83, 89, 91, 92, 94,
    97, 98
  flight-or-fight brain chemicals and
    enhanced memory, 36
  glutamate activity in, 104, 196
  hippocampus, 110, 157
  hypothalamus, 89–90, 94, 140, 157
  limbic system, 24
  lissencephaly (smooth brain), 207
  medical commonalities between
    humans and animals, 49–50, 197

NMDA receptors, 71, 77
  rabies and, 205
  serotonin levels in, 78, 144, 145, 245,
    246, 247
  temporal lobe epilepsy or seizures, 5
  (*see also* seizure disorders)
Brainstorms: Epilepsy in Our Words
  (Schachter), 98
Briand, Servane, 222
Brock (dog with bizarre aggression), 4–5
Brody (dog exhibiting jealousy), 52
Brudnick Neuropsychiatric Research
  Institute, 75
Buddha, 57
Buddhist prayer of compassion, 173
Buddie (dog with panic attacks), 206–7
bulimia, 66
bulldogs, 134
Bull Terrier, The (Montgomery), 84
bull terriers
  about the breed, 84
  author's research on the breed and tail
    chasing, 82–85
  "bull terrier syndrome," 83, 87
  canine autism and, 85–88, 260
  diet and, 88
  explosive aggression in, 2, 82, 85
  extreme fear reactions in, 90–91
  Fragile X syndrome and, 86
  luteolin treatment for, 88
  phobias in, 82, 84
  prey drive in, 84
  seizure disorders in, 76, 83, 97
  tail chasing, 75, 82–83, 84, 97, 98
  "trancing," 84, 85
  zinc deficiencies in, 76
Bully (dog with bizarre aggression), 2
Burghardt, Walter, 39–40, 203
Burmese cats, wool sucking by, 75
Buspar (buspirone), 123, 139, 146, 147,
  149, 164–65, 216
butorphanol, 208, 216

C

Calgary Zoo, 65
cancer
  brain tumor, 205–7
  leading medical cause of dog death, 212

nocturnal anxiety and, 201, 216–18
stress and, 212
canine cognitive dysfunction (CCD),
187–90, 195, 196, 197, 198
canine compulsive disorder, 64
canine epilepsy. *See* epilepsy
canine heart problems, 211
car (or bicycle) chasing, 141, 150
Case Western Reserve University, 197
Cassie (cow with agoraphobia), 182–83
catecholamines, 224, 246
catenins, 78–79
cats. *See also specific behaviors; specific cats*
aggression in, 95, 96, 127, 129–30,
148, 231
alpha cat syndrome, 205–6
balance and, 261
brain tumor in, 205–6
compulsive behaviors in, 59, 64–65,
67–68, 75
depression in, 238
displaced suckling behaviors, 64–65
experiments on thought processes in,
257–58
"feline affective defense display," 127
feline urine marking, 122–24, 161,
165
FHS in, 96
grieving by, 233–34
hair stripping as OCD, 24, 59, 68
hiding behavior, 162
hypothalamus and predatory behavior,
90
introducing a new cat to another,
133
kinesthetic intelligence of, 261
medications for aggression, 148
misdirected aggression in, 129–30
nocturnal anxiety in, 215
as "obligate groomers," 67
pilling a cat, suggestion, 231
"piloerection," 127
prey drive in, 140, 142
seizure disorders in, 95–96
senility in, 191–92
separation anxiety and, 161
social anxiety and, 162
socialization of, 162

territoriality in, 133, 148
thyroid problems in, 201, 230
treatment for compulsive behavior,
59, 68
unneutered males, appearance of, 123
cats, big (wild species), 65. *See also*
cheetahs
Celexa (citalopram), 248
Central Park Zoo, NYC, 65–66
cheetahs (*Acinonyx jubatus*), 195, 197
Chesapeake Bay retrievers
bizarre aggression in, 4–5
compulsive swimming in, 63
chimpanzees, 254–55, 256
Chirac, Jacques, 241–43
chlorpromazine, 247
Clomicalm, 200
clomipramine, 60, 66, 145, 165
clonazepam, 2, 214
clonidine, 34–35, 151, 246–47
Comet (dog with bizarre aggression), 1–2,
214
community of all life, 243
CoQ10, 190
Corson, Samuel, 202
cortisol, 157
cowering, 32, 42, 162, 164
cows, 44, 166, 182–83
Cassie, agoraphobia and, 182–83
crating dogs, 53–54, 160, 239, 250
crate panic, 160–61
cribbing (in horses), 16–22, 25–26, 58,
167
arcane and cruel treatments for, 17
author's study of, 18
endorphins and, 22
ethological approach to, 61
health consequences of, 17, 18
medication for, 19–20, 22, 23, 59,
71–72
stress of confinement and, 17, 20–21
study of drug treatment for, 25–26

D

Dalai Lama, 115
Daniel E. Holland Military Working Dog
Hospital, Lackland Air Force Base,
39

Dan the Man (quarterhorse with flank biting behavior), 104–6, 110
Darwin, Charles, 27, 257, 258, 260
deer, 155
defecation (pooping)
  "accidents" and senility, 187, 189, 190, 191
  "accidents" and separation anxiety, 42, 160
  of dogs, in the house, 42
depression, 11, 144, 233–48
  aggression and, 241–42
  in cats, 238
  commonality in animals and humans, 233
  in dogs, 238–40
  dysthymic disorder, 238–39, 241
  grief and, 233–38
  learned helplessness and, 239–40
  liver failure and, 204
  medications, 145, 148, 234, 242, 247–48
  antidepressants, 109–10, 139, 143, 144–45, 149, 165, 188, 200, 221, 222, 247–48, 234 (see also antidepressants; Prozac)
  personality traits and predisposition to, 240–43
  physical disorders linked to, 211, 234
  serotonin and, 144, 224
  "shutdown" and, 240
  symptoms, 238
  thyroid problems and, 224
  treatment, nondrug, 234, 235
Descartes, René, 10, 12–13, 209
desipramine, 149
dextromethorphan, 72, 73, 80
diabetes, type 2, 211
Diagnostic and Statistical Manual (DSM), 33n, 150
  diagnosis of GAD, 163
  OCD as anxiety-type disorder, 69
  PTSD classification in, 33
Dick, Philip K., 220
digging, 63
Do Androids Dream of Electric Sheep? (Dick), 220

Doberman pinschers
  aggression treated with dopamine blocker, 149
  compulsive behaviors in, 62, 70, 76–78, 260
  flank-sucking, 62, 75, 76
  genetic component of compulsive behaviors in, 76, 77–78, 87
  rescue Dobie and Beanie Baby toys, 76
  "shopping" by, 76
  stress and, 70
Dodman, Gwendoline "Gwen," 44–48, 198
  dementia and, 192–94, 198
  parrot of, 258
Dodman, Nicholas H. See also One Medicine; specific animal problems
  arrives at Tufts, 51
  bitten by dog, 126–27
  borderline hypothyroidism study, 229
  cats of, Monkey and Cinder, 129–30, 191–92
  childhood and upbringing, 44–48
  controlled squeezing experiment, 168–71
  daughter Keisha, 53–54, 161–62, 181
  as director, Animal Behavior Clinic, Tufts Cummings School of Veterinary Medicine, 11
  empathy and affection for animals, 48
  "End-of-the-Line Dodman," 221
  first behavioral condition investigated, 18
  Griswald the cat and, 41, 54
  horses of, Dolly and Jack, 236–38
  interview on PTSD in dogs, 31
  methodology of, 22
  mother's Alzheimer's disease and, 192–94, 198
  motto for behavioral pharmacology, 151
  OCD's genetic link study, 75–80
  One Medicine and, 23, 51
  opioid antagonists for compulsive behavior study, 71
  parrot of, 258

Pippit (song thrush) and, 44–46, 258
as a professor of veterinary science
   specializing in animal behavior, 2
pushback from fellow scientists, 8–9
rescue dog Jasper and, 53–56
rescue dog Rusty and, 41–43, 126
research on tail chasing in bull terriers,
   82–85
sister Angela, 140, 175–76, 192
study on compulsive behaviors in dogs,
   58–59
study on reverse tolerance, 21
TSA funds Tourette's study in horses
   and, 108
veterinary anesthesia and, 49–50, 51
wife Linda, 42, 43, 231
work on NMDA blockers, 73
dog bites, 122, 125–27
anxiety-based territorial aggression and,
   135
"fear biters," 134–35
misdirected aggression and, 130
story of Ruckus, 249–52
territoriality and, 134
dog fights, 128–29, 130, 133
dogs. *See also* ADHD; aggression;
   Alzheimer's disease; anxiety;
   autism; depression; obsessive
   compulsive disorder; phobias;
   PTSD; seizure disorders; sleep
   problems; thyroid problems;
   *specific behaviors; specific breeds;
   specific dogs*
acceptance of new concepts about, 6
altruism in, 9–10
body language and, 261
canine predatory behaviors, 61–62,
   89–90
castration of, 122
diagnosing strange behavior, by looking
   for parallel human behaviors, 6,
   8–9
"dog years" formula, 190
as evolutionary relative to humans, 3,
   13
exercise needed daily, 117, 128
as experimental models in psychological
   and neuroscience experiments, 51

experiments on thought processes in,
   257
grieving by, 234–36
human medications for, 2–3 (*see also*
   medications; One Medicine)
ineffective training methods, 43
jealousy in, 51–52, 157
laughter by, 157
number in U.S., 58
number with separation anxiety, 159
owner's emotional states mirrored in,
   118
predatory response, 137–40
protein in diet and, 118, 128, 131,
   204
sadness or joy in, 157
self-grooming in, 62–63
senility in, 187–90
sibling rivalry in, 132
submissive behaviors, 42, 162–63
survival-linked behaviors, 62–63
territoriality in, 133–35
top dog and pack hierarchy, 131,
   132–33
in war zones, sensory assaults on, 32
dog training, 81–82, 128
behavioral modification, 165
counterconditioning, 136, 180, 250
dealing with predatory aggression,
   138–40
desensitization, 35, 136, 180, 181,
   185
for dog with fear aggression, 150–52
front doors and foyers, territorial
   aggression and, 135–36
head halter for, 141, 150–51
leadership status and, 131, 134, 135
socialization, 127–28
stopping car chasing, 141–42
Dog Who Rescues Cats, The, 9–10
Dolly (horse who loses her soul mate),
   236–38
dolphins, 256
dopamine, 148, 188, 224, 247
Doust, Bill, 27–30
dreaming, 220
Drews, Marilyn, 84
Dumas, Alexandre, 248

E

Edwards, Emily "Mimi," 99–104, 109
Elavil (amitriptyline), 139, 145, 203, 221, 234
elephants, compulsive behaviors in, 65
Eli Lilly company, 249
Elsa (dog with PTSD), 27–30, 33, 34
Emma (beagle with ADHD), 199–201
Emory University, PTSD study at, 38–39
emotional problems, 41–56
   Jasper (neglected rescue dog), 53–56
   Rusty (fearful rescue dog), 41–43
empathy, 7, 9, 13, 44, 47, 48, 111, 166, 254
endorphins, 20–21, 22, 71, 104, 157, 167–68, 171
   hugging and, 171
English bulldogs, 218–19
epidural morphine, 208
epilepsy (canine epilepsy), 5, 64, 79–80, 83, 92, 93, 97–98. *See also* seizure disorders
epinephrine, 157
ethology, 61

F

fear, 90–92, 153–71, 175. *See also* anxiety; fear aggression; phobias
   anxiety vs., 154–55
   brain and, 157
   dog with extreme fear reaction, 90–91, 98
   epilepsy and, 98
   "fear learning" and, 155
   genetically determined fears, 155
   in German short-haired pointer, 147
   medications for, 147
   "pharmacologic desensitization," 147
   situational, 154–55
fear aggression
   behavioral modification with medication for, 150–52
   in cats, 127
   in dogs, 115–19, 125–27, 150–52, 254
   fear-based territorial aggression, 135
   medications for, 246

preventing, 127–28
   triggers of, 150
fear responses, 155
   amygdala and, 36, 38, 157
   automatic internal responses, 156
   fight or flight, 155
   freezing in position, 155
   hiding, 155
   hippocampus and, 157
   salivation in dogs, 156, 157
   seeking protection, 155
   thigmotaxic behavior, 155
feline cognitive dysfunction (FCD), 191–92, 196, 197
feline hyperesthesia syndrome (FHS), 96
fence-walking, 65
Ferrero, "Big John," 112, 113
Ferrero, Willie, 110–13
firework phobia, 180
flank sucking, 62, 75, 87
Fleishman, Bob, 68
fly snapping, 62, 89, 90, 92
Fossey, Dian, 47
Four *Es*, 253–54
Fragile X syndrome, 86, 260
Franklin Park Zoo, Boston, 66

G

Galdikas, Biruté, 47
generalized anxiety disorder (GAD), 163–64, 171
genetics
   aggression in male Malinois and, 143
   anomaly in bull terriers, 86
   autism and, 87
   cadherin family of genes, 77, 87, 113, 260
   compulsive behaviors and, 75–80, 87, 260
   Doberman pinschers and compulsive behaviors, 76, 77–78, 87
   FoxP2 for speech and language, 261
   Illumina "chip" gene mapping, 86
   narcolepsy and, 221
   phobias and, 176–77
   PTSD and, 35, 36, 37–38, 39, 40, 77
   relin gene and lissencephaly, 207
   research on inherited behaviors of animals, 81

SLITRK1 gene, 113
social anxiety and, 162
tail chasing and, 82–83
Tourette's syndrome and, 102–3, 104, 113
various fears determined by, 155
German shepherds
ADHD in, 202
compulsive behaviors in, 62
hypothyroidism in, 227–28
prey drive in, 140
PTSD in war dogs, 30–34
seizures in, 88, 91–92
tail chasing, 63–64, 88
German short-haired pointer, 147
Gina (Army dog with PTSD), 30–33, 34
Ginns, Edward, 75, 78, 85
Glen, Ian, 243
glucosamine-chondroitin, 74
"glugger" or "snoofer" (dogs with seizures), 93–94
glutamine, 71
Golden retrievers, 1–3, 48, 214, 260
Goodall, Jane, 47, 254–55, 258
gorillas, 48, 255, 256
OCD-like condition in, 66
Grandin, Temple, 165–71, 258
Graves' disease, 230
Green Jungle Chicken (author's parrot), 258
Greyfriars Bobby (legendary dog), 235–36
grieving animals, 233–38
Gus (polar bear and compulsive swimming), 65–66

H
Hachiko (legendary dog), 236
Haggerty, Captain Arthur, 81
Haldol (haloperidol), 111, 148
Handler, Scott, 109
Hare, Brian, 51
Harris, Christine, 52
Hauser, Marc, 9
Haynes, Eric, 31
Heath, Terry, 88
Herriot, James, 153
Hill's Prescription Diet b/d, 189–90

horses
"bowing a tendon," 111
compulsive behaviors in, 16–22, 24–26, 58, 65
cribbing, 16–22, 25–26, 65, 75
flank biting, 99–113
flank biting survey, 105–7
grieving by, 236–38
jugular vein, 19
of Mongol warriors, 19
refusal to cross thresholds, 102, 107, 111
"sand colic," 105
separate visual fields of, 261
stress of confinement, 17, 20–21
study of endorphins in, 22
Tourette's syndrome in, 99–113
in the wild, eating behavior, 17
"wind-sucking," 16, 20, 25
Huntington's disease, 92
huperzine, 91–92, 93, 196–97
hyperactivity, 8. See also ADHD
hypervigilance, 29, 32, 33, 40, 156, 162, 174
hypothyroidism or hyperthyroidism. See thyroid problems

I
imipramine, 247
Imperial Chemical Industries (ICI), 243–44
India's Ministry of Environment and Forests, 256
instrumental aggression, 131–33
Irish setters, 207
irritable bowel syndrome (IBS), 211, 225
isoflurane, 244

J
Jackson (dog with ADHD), 202
Jackson, Michael, 244
Jake (dog with night terrors), 213–15
Jasper (neglected rescue dog), 53–56, 239
jealousy, 7, 51, 241
in dogs, 51–52, 157
Jenike, Michael, 73, 74
Johnson, Samuel, 106
Journal of the American Veterinary Medical Association, 228

K

Kalie (cat with psychogenic alopecia), 68
ketamine (Special K), 71–72, 74–75
Kilcommons, Brian, 81–82
King (gorilla with great memory), 255
King Charles spaniel (nonstop chasing behavior), 73–74
Klonopin (clonazepam), 147
Korowai tribe, 69
Kundera, Milan, 137

L

Labrador retrievers
  phobia in, 173–75
  PTSD in, 40
Leakey, Louis, 255
Levine, Mark, 112
levothyroxine, 224, 226
Lhasa apsos, 207
lick granuloma, 57–58, 60. See also acral licking (acral lick dermatitis)
Lief (dog with compulsive tail chasing), 63–64, 89
Lillian (angry cat), 95–96
Lincoln, Abraham, 250
lissencephaly (smooth brain), 207
listlessness, thyroid problems and, 201, 225
Lita (dog exhibiting extreme fear behavior), 147
liver problems, 203–5
  in children, 204
  failure, 203–4, 234
  portosystemic shunt, 204–5
L-thyroxine, 229
Lucky (dog with fear aggression), 150–52
Luescher, Andrew, 202–3
Luvox (fluvoxamine), 248

M

Mabel (black Lab with insect phobia), 173–75
male and intermale aggression, 121–24
Malinois (short-haired Belgian shepherd), 143
Maltese terriers, 204–5, 241–42
MAOI (monoamine oxidase inhibitor), 188

Maple Farm Sanctuary, 183
Marty (dog with canine Alzheimer's), 187–90
maternal aggression, 124–25
Maxwell (depressed cat), 233–34
McConnell, Patricia, 81
Mearns, Hughes, 62
mechanistic theory, 7, 10, 12–13, 156, 209. See also Descartes, René; Morgan's Canon
medical marijuana for pets, 136
medications. See also specific drugs
  for ADHD, 200–201, 203
  for aggressive behaviors, 139, 143–52, 246
  for Alzheimer's disease, 188–89, 195–97
  anesthetics, 244
  animal testing and, 244–48
  antipsychotics, 111, 148–49, 247
  for anxiety, 8, 66, 83, 92, 136, 162, 164–65, 175
  author's joint study of "reverse tolerance," 21
  beta-blockers, 36–37, 175
  for compulsive behaviors, 58–59, 60, 64, 65, 66, 68, 71–75, 79, 210
  for cribbing, 20, 21, 25–26
  for depression, 109–10, 139, 143, 144–45, 149, 165, 188, 200, 221, 222, 247–48, 234
  for depression, side effects, 145–46
  drugs of addiction, 18, 21
  medical marijuana for pets, 136
  for narcolepsy, 221, 222
  for nocturnal anxiety, 218
  opioid antagonists, 20, 21, 22, 68, 71, 75, 100, 101–2, 108, 168, 169, 210
  for pain, 208
  for phobias, 175, 181, 184, 185
  placebo, use of, 20, 229
  for PTSD, 34–35, 36–37
  for seizure disorders, 2–3, 5, 6, 8, 64, 83, 88, 89, 91, 92, 95, 96, 98
  for separation anxiety, 162, 165, 200
  for thyroid problems, 224, 226, 229, 231
  for Tourette's syndrome, 101–2, 103, 108, 109–10

treating animals with human meds, 3, 8, 12, 13–14, 20, 25–26, 150, 165, 242, 243–45, 248, 259–60 (*see also specific disorders; specific drugs*)
melatonin, 3, 190
Meyer, Jean, 51
mice
  compulsive behaviors in cages, 70–71
  drug testing and, 244–48
  fear of, 176
  medications for compulsive behavior, 72
Miczek, Klaus, 5
Migdol (Arabian stallion with flank biting), 99–104, 109–10, 113
Mignot, Emmanuel, 221–22
Milne, A. A., 1
Mindy (dog with stranger anxiety), 153–54, 162
misdirected aggression (redirected aggression; irritable aggression), 128–31
  how to deal with, 129–30
  in people, 130
Mobey (horse exhibiting cribbing), 24–26, 58, 61, 71
Monaco, Anthony, 87
Monkey (cat with feline Alzheimer's disease), 191–92
Monkey and Cinder (misdirected aggression in cats), 129
monkeys, 146
  serotonin experiment with, 245–46
  skin picking in, 210
Montgomery, E. S., 84
Moon-Fanelli, Alice, 84
Morgan, C. Lloyd, 10, 156
Morgan's Canon, 10–11, 13, 16, 156, 165, 259
Moyer, Kenneth E., 121

N

nalmefene, 101–2
naloxone (Narcon), 20, 21, 22, 71, 101
naltrexone, 168, 169, 210
Namenda (memantine), 73, 74, 79, 195–96
narcolepsy, 220–22
  in dogs, 221–22
  genetic component, 221
  in horses, 221
  treatment, 221, 222
  triggers of, 221, 222
National Institutes of Mental Health (NIMH), 18, 77, 86
Neff, Mark, 78
neural cadherin (CDH2), 77, 113
neuroleptic drugs, 247
neurotensin, 87, 260
neuroticism, 241
Newfoundlands, compulsive swimming in, 63
nitromemantine, 196
NMDA blockers, 71–75, 88, 195–96
noise phobias, 178, 206, 215, 216
Nonhuman Rights Project, 256
norepinephrine, 145, 147, 151, 175, 224, 246
Normile, Jo Anne, 104–6, 110

O

Obama, Barack, 63
obsessive-compulsive disorder (OCD). *See also specific animals; specific behaviors*
  absent in wild animals, 69
  anorexia nervosa in humans as, 210
  anxiety and, 61, 69, 154, 164, 210
  author's grant denied because of outmoded view of animals, 8–9
  brain and, 24, 75, 260
  bulimia as, 66
  canine breeds prone to, 62, 63, 70
  catenins and, 78–79
  in cats, 59, 64–65
  CDH2 (gene) and, 77, 78
  commonality in people and animals, 75, 77
  compulsive behaviors in animals, 8, 16–22, 24, 57–80
  defined, 23
  in dogs, 58–64, 70, 76–78, 79–80
  epilepsy linked to, 64, 79–80
  examples of, in humans, 23–24, 67, 78
  genetic component, 75–79, 82–83, 260
  in horses, 16–22, 24–26, 58, 65 (*see also cribbing*)

obsessive-compulsive disorder (OCD)
(*cont.*)
  medications to treat, 58–59, 60, 64,
    65, 66, 68, 71–75, 78, 79, 195
  NMDA blocker to treat, 72–73,
    259–60
  overgrooming and, 68–69
  Rapoport study about childhood
    obsessive-compulsive disorder,
    60–61
  smoking and drinking as, 72–73
  stress triggering, 70
  Tourette's syndrome and, 77, 113
  VBM testing and, 78, 197
  in zoo animals, 65–66
Ogata, Niwako, 78
Oliver (depressed dog), 234–35
One Medicine, 8, 13, 14, 18, 165,
  192–94, 198, 217–18, 253
  author's commitment to, 23
  development of drugs for use in
    humans and animals, 74
  diagnosing behavior in pets, by looking
    for parallel human behaviors, 8–9,
    23
  medical commonalities between
    humans and animals, 26, 38, 49–51,
    229
  Meyer and, 51
  opponents to, 26
  PTSD study and, 38
opioid or morphine antagonists, 20, 21,
  22, 68, 71, 75, 100, 101–2, 108,
  168, 169, 210
Oprl1 gene, 38, 39
Ostrander, Elaine, 86
oxytocin, 118–19
Ozzy the Oddball (dog exhibiting
  compulsive behaviors), 79–80

P

pacing behavior, 65, 156, 157, 158, 160,
  187, 189, 204
pain
  anxiety and, 208
  behavioral problems and, 208–9
  sleep problems and, 217–18
  treatment, 208

panic attack or disorder, 34–35, 158–59,
  175
  agoraphobia and, 159
  brain tumor causing, 206–7
  separation anxiety and, 159
parrots, 255
  feather-plucking, 66–67
  intelligence of, 66, 258
Pavlov, Ivan, 51
Paxil, 165
Pekingese, 218
Pepper Belle (Standardbred mare with
  Tourette's syndrome), 110–13
Pepperberg, Irene, 66
phenobarbital, 2, 5, 83, 91, 94, 96
  alternative to, 92
  side effects, 92
phobias, 82, 156, 164, 173–85
  agoraphobia, 159, 176, 182–85
  "animal type," 173–76
  author's daughter, fear of needles, 181
  author's sister, wasp phobia of, 175–76
  blood/injection/ injury type, 181
  of car travel, 181
  counterconditioning for, 180
  defined, 174, 176
  desensitization for, 180, 181, 185
  fear of bridges, 177
  fear of fireworks, 180
  fear of noises, 178, 206, 215, 216
  fear of thunderstorms (storm-phobic),
    171, 176–81
  genetic component, 176–77
  medication for, 175, 181, 184, 185
  natural environmental-specific, 159
  "other type," 181–82
  pressure vest for, 179–80, 181
  situational type, 181
  treatment (nondrug), 174–75, 179–80
pica, 64–65
pigs, 166–67
  controlled squeezing experiment,
    168–71
Pippit (rescued song thrush), 44–46, 258
pitbull terrier. *See also* bull terriers
  anxiety-based territorial aggression, 135
Plato, 223
Poker's Queen Bee (horse exhibiting

cribbing), 16–22, 25, 58, 61, 71
polar bears, 65–66
Polly (parrot of Gwen Dodman), 258
Portuguese fishing dogs, compulsive
    swimming in, 63
post-traumatic stress disorder (PTSD),
    27–40
  automobile accidents and, 34
  brood bitches in puppy mills and, 34
  C-PTSD (canine PTSD), 39–40
  defined in DSM, 33, 33n
  desensitizing retraining actions, 35
  diagnosis of, 33–34, 39–40
  in dog following painful treatment at a
    veterinary hospital, 34–35
  in dog shot by police officer, 27–30,
    33, 34
  in dogs, triggered by smell of lamb
    cooking, 35
  fear-based learning and, 36
  genetic predisposition, 35, 36, 37–38,
    39, 40, 77
  Labrador retrievers and, 40
  media story about PTSD in dogs, 31
  medications for, 34–35, 36–37
  Oprl1 gene and, 38, 39
  in people, 33, 35, 37
  research in lab animals, 36
  sleep problems and, 216
  symptoms, 32–33, 40
  treatment, 30, 33, 37–38
  war dogs and, 30–33, 34, 35, 37, 38,
    39–40
potassium bromide, 91
predation, 89–90, 137–42
  appetitive behavior and, 61–62, 89–90,
    141
  chasing cars, running children, joggers,
    bicyclists, or skateboardists and,
    141–42
  in cats, 140, 142
  consummatory phase compulsions
    and, 62
  dog breeds with strong prey drive, 84,
    90, 140–41
  exercise to curtail, 142
  head halter for preventing car chasing
    training, 141–42

high prey drive that "dyslexes"
    (erroneous fixation on nonprey
    subjects), 141
predatory aggression
  infant seen as prey, 137–40, 141
  treatment, 138–40
Propofol, 244
propranolol, 36, 147
Provigil, 222
Prozac (fluoxetine), 2–3, 60, 65, 73–74,
    78, 88, 89, 96, 123, 146, 151, 152,
    165, 246, 248, 250, 251
psychogenic alopecia, 67
psychogenic anorexia, 160
psychopharmacology, 247
psychosomatic conditions, 209–12
  acral lick dermatitis as, 210, 225
  medical conditions associated with,
    209, 225
  medications for, 210
  neurotic excoriation, 210
  skin picking or feather picking, 210
PTSD. See post-traumatic stress disorder
puppy mills, 34, 162, 163, 164, 239

R
rabies, 205
Rapoport, Judith, 60–61
Ratey, John, 149–50
redirected aggression. See misdirected
    aggression
Reisner, Ilana, 4
rescue dogs, 41–56
  author's Rusty and Jasper, 41–43, 47,
    53–56
  Doberman and Beanie Baby toys, 76
  fear-aggression in, 115–19
  learned helplessness in, 239
  separation anxiety in, 28
resilience, 35–36, 53
Ritalin, 200–201
rock chewing, 62
rodents, 155. See also mice
Rottweiler, aggressive behavior and
    clomipramine treatment, 145–46
Ruckus (dog with serious aggression),
    249–52
Rusty (fearful rescue dog), 41–43, 47

S

Sacks, Oliver, ix, 148, 167
salivation in dogs, 51, 88, 91, 145, 156, 157
samoyeds, 207
Samson and Delilah (dogs hostile to new baby), 137–40
scent hounds, 142
Schachter, Steve, 92, 97–98
schnoodle (schnauzer-poodle mix), 79–80
Schopenhauer, Arthur, 41
Schweitzer, Albert, 249
seizure disorders, 2, 8
    aberrant behaviors and, 93–94
    aggressive behavior and, 2–3, 4, 5, 88, 94, 95–96
    autism and, 85, 88
    "bull terrier syndrome" and, 83, 88
    in cats, 95–96
    caused by medication, 145–46
    compulsive behaviors and, 64, 79–80, 88–98
    diagnostic testing for, 5
    EEGs for animals and, 5, 83, 89, 91, 92, 94, 97–98
    epilepsy or canine epilepsy, 64, 79–80, 83, 92, 93, 97–98
    "episodic dyscontrol" in humans, 5, 6, 95
    in German shepherds, 88
    "glugger" or "snoofer" dogs, 93–94
    grand mal, 4, 88, 89, 94–95
    medications for, 2–3, 5, 6, 8, 64, 83, 88, 89, 91, 92, 95, 96, 98
    nocturnal seizures, 3
    partial seizures, 4, 5, 64, 87–98, 225
    partial seizures, symptoms of, 94–95
    postictal period, 94
    self-injurious behaviors and, 96–97
    sleep and, 98
    status epilepticus, 83
    temporal lobe epilepsy, 5
    triggers of, 91, 98
selegiline (Anipryl, Eldepryl), 188–89
separation anxiety, 27–30, 42, 159–62, 241
    in cats, 161
    in children, 161–62
    in dogs, behaviors associated with, 160, 199
    medications for, 162, 165, 200
    number of U.S. dogs with, 159
    prevention, 162
    questions for dog owners, 159
    storm phobia and, 179
    training program for, 200
    treatment, 162, 171
serotonin, 78, 139, 144–52, 222, 224
    depression and, 144, 247–48
    link to aggressive behavior, 144, 245–46, 260
    low levels, symptoms, 224
    as a "neuromodulator," 144
    suicide and, 144
shelter dogs, 153. See also rescue dogs
    separation anxiety and, 159
shock collars, 250–51
Shuster, Louis, 18, 21, 22, 25–26, 58–59, 71, 72–73, 100, 108, 112, 113, 194
"shutdown," 240
Siamese cats
    seizures and tail biting in, 96
    wool sucking by, 75
sight hounds, 142
Sin-Horng, Jong, 248
Skinner, B. F., 51
slaughterhouses, 166–67
sleep apnea, 202, 218–19
    continuous positive airway pressure (CPAP) for, 219
    dog breeds prone to, 218
    symptoms, 219
    treatment, 219
sleep problems, 28–30, 33, 213–22
    medical issues and, 216–18
    medications for, 218
    narcolepsy, 220–22
    night terrors, 213–15
    nocturnal anxiety, 215, 216–18
    nocturnal anxiety in children, 218
    noise phobias and, 216
    pain and, 217–18
    PTSD and, 216

RBD or REM Behavior Disorder, 213–15, 220
serotonin and, 224
Sundowner Syndrome, 215–16
Smokey (cat and urine marking), 122–24
snakes
fear of, 155, 176
venom milking, 60
Snowball (pacing polar bear), 65
social anxiety, 162–65
in children, 163
socialization, 117, 127–28, 153, 162
spiders, fear of, 155, 186
Spike (cat with tail biting behavior), 96
Spock (dog with panic attacks), 158
sporting breeds, 142
springer spaniels, 141
predatory aggression in, 137–40
rage and, 98
SSRIs (selective serotonin reuptake inhibitors), 165, 247–48. See also Prozac
stall-walking, 65
Stanford Center for Sleep Sciences and Medicine, 221
Star (dog with panic reaction due to PTSD), 34–35
Stella (dog with extreme fear reaction), 90–91, 98
stereotypies, 59, 60, 100
in caged mice, 70–71
reclassified as compulsive disorders, 61
tail chasing and, 82, 83
stranger anxiety, 153–54, 254
in children, 163
in puppies and kittens, 163
stress. See also post-traumatic stress disorder (PTSD)
alleviating, 171
appetite and, 210
beta-blockers to prevent high-stress learning, 36
bulimic gorilla and, 66
cardiovascular function and, 210–11
compulsive behaviors in animals and, 61
cribbing (in horses) as reaction to, 20

diabetes and, 211
IBS and, 211
immune system and, 211–12
owner's, transmitted to pet, 118
PTSD and, 36
in puppies, separation anxiety and, 159
separation anxiety and defecation, 42, 160
skin conditions and, 209–10
Tourette's syndrome and, 104, 107
triggering animal compulsions, 70
stress hormones, 36, 118, 157, 158, 166, 210, 211, 240. See also specific hormones
Sumo (dog with depression), 241–43
systemic desensitization, 148

T
tail chasing, 24, 63–64, 75, 82–83, 88, 90, 98
medications for, 89
tail wagging, 35, 164
Talmud, 187
Tank (bulldog with territorial aggression), 134
Tapazole, 231
Terrell, Ross, 244
terriers, prey drive in, 140, 141, 142. See also specific terrier breeds
territorial aggression, 133–36
in cats, 148
front doors and foyers as battleground, 135–36
testosterone
aggression and, 121–22, 125
feline urine marking and, 122–24
Thalhammer, Johann, 5
Thatcher, Margaret, ix
Theoharides, Dr., 86–87
theory of mind, 51–52, 157
therapy dogs, 203
Thornton, Big Mama, 56
Three Musketeers, The (Dumas), 248
thunderstorm fears (storm phobia), 171, 176–81, 199
recommendations for, 179
separation anxiety and, 179

Storm Defender and Anxiety Wrap to
  treat, 179–80
typical dog behaviors and, 177–78
thyroid problems, 223–32
  aggression and, 227–28, 229, 230,
    231
  anxiety and, 226–27
  borderline hypothyroidism, 223–24,
    225, 227–28
  in cats, 230
  commonality in animals and humans,
    232
  hyperthyroidism, 230–32
  hypothyroidism, 224–25, 226, 229
  listlessness and, 201
  medications for, 224, 226, 229, 231
  in purebred dogs, 226, 227
  symptoms in pets, 225, 227, 229, 231
  thyroid tumors, 230, 231
Tiemer, Dick and Nancy, 249–51
Tourette's syndrome, 11, 77
  coexistent compulsive disorders, 107
  genetic component, 102–3, 104, 113
  hemiballismus and, 107
  in horses (Equine Tourette's
    Syndrome), 99–113
  in humans, 106–7
  medications for, 101–2, 103, 108, 109,
    111
  OCD and, 113
  Pepper Belle as mascot for the TSA,
    111–13
  stress and, 104, 107
Tourette's Syndrome Association (TSA),
  108, 111–12
tramadol, 93
trazodone, 143
trichotillomania, 67, 68
  anxiety and, 69–70
  beard plucker, 70
  eyebrow plucker, 70
  feline form of, psychogenic alopecia, 67
Tufts University, 51, 87, 226–27
  Medical School, 18, 21
  North Grafton Campus, 81
  patent on NMDA receptor blockers,
    72, 74

Tufts University, Cummings School of
  Veterinary Medicine, 18–19
  Animal Behavior Clinic, 11, 242
  behavior clinic, 226–27
  Dodman at, 51, 242
  experiments on pain relief in dogs, 208
  Internal Medicine Department, 234
  Large Animal Hospital, 19
  Mobey (horse exhibiting cribbing) and,
    24–26
  PTSD as canine condition and, 30
Twain, Mark, 230
20/20 (TV show), 149

U
"underdog" or "top dog," 132–33
University of California, Davis veterinary
  school, 143
University of St. Andrews, Scotland,
  256
urination (peeing)
  "accidents" and senility, 187, 189,
    190
  "accidents" and separation anxiety,
    160
  canine urine marking, 134
  feline urine marking, 122–24, 161,
    165
  squat-and-pee greeting, 199
  submissive urination, 42, 162–63

V
Valium (diazepam), 66, 83, 92, 143–44,
  147, 149
Van Andel Institute, 78
veterinary profession, 12–13
  behavioral medicine and, 12, 13, 16
  DAMN-IT acronym, 225–26
  observation as means of diagnosis, 16,
    224
  "treating" people and, 99
voxel-based morphometry (VBM), 78, 197
V-troughs, 166–67, 168

W
Walker hound mix, 150–52
weaving (in horses and zoo animals), 65

whales, 259
Wheaten terrier, soft-coated, 249–52
wirehaired fox terriers, 207
Wise, Steven M., 256
Wong, David T., 248
Woodhouse, Barbara, 81
wool sucking, 62, 64–65

X

Xanax (alprazolam), 147

Y

Yorkshire terriers, jealousy and, 52
YouTube videos
    demonstrating empathy in dogs, 9
    dog with sleep disorder, 214

Z

Zoloft, 60
zoo animals, compulsive behaviors in,
    65–66. *See also specific animals*

About the Author

Dr. Nicholas Dodman, BVMS, DVA, is a Diplomate of the American College of Veterinary Behaviorists, professor emeritus at Tufts University, and former director of the Animal Behavior Clinic at the Cummings School of Veterinary Medicine. One of the world's most noted and celebrated veterinary behaviorists, he grew up in England and trained to be a vet in Scotland. At the age of twenty-six, he became the youngest veterinary faculty member in Britain. In 1981, Dr. Dodman immigrated to the United States, where he became a faculty member of Tufts University School of Veterinary Medicine. There, he founded the Animal Behavior Clinic—one of the first of its kind—at Tufts in 1986. Dr. Dodman has written four acclaimed bestselling books, *The Dog Who Loved Too Much, The Cat Who Cried for Help, Dogs Behaving Badly,* and *If Only They Could Speak.*

Dr. Dodman has also authored two textbooks and more than a hundred articles and contributions to scientific books and journals. He is a member of the American Veterinary Medical Association and is a Leadership Council Member of the Humane Society Veterinary Medical Association. Dr. Dodman holds several US patents for behavior modification treatments, including a 2002 patent that details a novel treatment for obsessive-compulsive disorder in humans. Work in the Harvard and Yale University Psychiatry Departments confirms the validity of this novel treatment. He is married to Dr. Linda Breitman, a veterinarian who specializes in small animals. They and their children live in North Grafton, Massachusetts.